A PRIMER OF
SIGNAL DETECTION THEORY

D. McNicol

Professor Emeritus, University of Tasmania, Australia

With a New Foreword by Brian C. J. Moore

 **Psychology Press**
Taylor & Francis Group

Originally published 1972.

First published by
Lawrence Erlbaum Associates, Inc. Publishers
10 Industrial Avenue
Mahwah, New Jersey 07430

This edition published 2012 by Routledge

Routledge
Taylor & Francis Group
711 Third Avenue
New York, NY 10017

Routledge
Taylor & Francis Group
27 Church Street, Hove
East Sussex BN3 2FA

Library of Congress Cataloging-in-Publication Data

McNicol, D.
    A primer of signal detection theory / D. McNicol.
        p.   cm.
Originally published: London : Allen & Unwin, 1972. With new foreword.
Includes bibliographical references and index.
ISBN 0-8058-5323-5 (pbk. : alk. paper)
1. Signal detection (Psychology) 2. Psychometrics.   I. Title.
BF441.M22   2004
152.8—dc22   2004056290

Foreword

Signal Detection Theory (SDT) has had, and continues to have, an enormous impact on many branches of psychology. Although its initial applications were in the interpretation of sensory processes, its domain has since widened considerably. For example, concepts derived from SDT are widely used in memory research and in studies of the processing of verbal information. SDT has been called by many a revolution and I do not think that is an exaggeration. A basic understanding of SDT has become essential for anyone with a serious interest in experimental psychology.

The classic work on SDT is *Signal Detection Theory and Psychophysics* by D. M. Green and J. A. Swets, originally published by Wiley in 1966 and reprinted with corrections by Kreiger in 1974. This remains a very useful source for advanced researchers and those with mathematical sophistication. However, for many readers, the descriptions and derivations will be beyond their grasp. A more recent and more user-friendly text is *Detection Theory: A User's Guide* by N. A. Macmillan and C. D. Creelman, originally published in 1991. The second edition of this book has just been published by Lawrence Erlbaum Associates. The Macmillan and Creelman book still assumes a good deal of mathematical and statistical sophistication, but it makes more use of visual analogies, and is intended especially as a practical guide to those actively involved in areas of research that depend on a good understanding and appreciation of SDT. In their preface, Macmillan and Creelman state "It could be the basic text in a one-semester graduate or upper level undergraduate course." However some undergraduates may find it too detailed if they simply want to get a basic understanding of SDT.

When I first had to teach SDT to undergraduates in psychology, I was delighted to come across *A Primer of Signal Detection Theory* by D. McNicol, published by George Allen and Unwin in 1972. This was the only text book I could find that covered SDT at an introductory level, and that assumed only limited skills in algebra and statistics. I used this book with success as my recommended text for several my two personal copies of the book both "went missing," and all copies in our library also mysteriously disappeared. I was left "in limbo" for many years. It is with relief and pleasure that I now write this foreword to a new printing of the book. I can strongly recommend the book as an introduction to SDT and its applications. It is suitable for use as a student text book, but will also be useful for teachers and researchers in psychology who need to acquire a working understanding of SDT.

—*Brian C. J. Moore, FmedSci, FRS*
Department of Experimental Psychology
University of Cambridge

Preface

There is hardly a field in psychology in which the effects of signal detection theory have not been felt. The authoritative work on the subject, Green's & Swets' *Signal Detection Theory and Psychophysics* (New York: Wiley) appeared in 1966, and is having a profound influence on method and theory in psychology. All this makes things exciting but rather difficult for undergraduate students and their teachers, because a complete course in psychology now requires an understanding of the concepts of signal detection theory, and many undergraduates have done no mathematics at university level. Their total mathematical skills consist of dim recollections of secondary school algebra coupled with an introductory course in statistics taken in conjunction with their studies in psychology. This book is intended to present the methods of signal detection theory to a person with such a mathematical background. It assumes a knowledge only of elementary algebra and elementary statistics. Symbols and terminology are kept as close as possible to those of Green & Swets (1966) so that the eventual and hoped for transfer to a more advanced text will be accomplished as easily as possible.

The book is best considered as being divided into two main sections, the first comprising Chapters 1 to 5, and the second, Chapters 6 to 8. The first section introduces the basic ideas of detection theory, and its fundamental measures. The aim is to enable the reader to be able to understand and compute these measures. The section ends with a detailed working through of a typical experiment and a discussion of some of the problems which can arise for the potential user of detection theory.

The second section considers three more advanced topics. The first of these, which is treated thoroughly elsewhere in the literature, is threshold theory. However, because this contender against signal detection theory has been so ubiquitous in the literature of experimental psychology, and so powerful in its influence both in the construction of theories and the design of experiments, it is discussed again. The second topic concerns the extension of detection theory, which customarily requires experiments involving recognition tests, to experiments using more open-ended procedures, such as recall; and the third topic is an examination of Thurstonian scaling procedures which extend signal detection theory in a number of useful ways.

An author needs the assistance of many people to produce his book, and I have been no exception. I am particularly beholden to David Ingleby, who, when he was working at the Medical Research Council Applied Psychology Unit, Cambridge, gave me much useful advice, and who was subsequently most generous in allowing me to read a number of his reports. The reader will notice frequent reference to his unpublished Ph.D. thesis from which I gained considerable help when writing Chapters 7 and 8 of this book. Many of my colleagues at Adelaide have helped me too, and I am grateful to Ted Nettelbeck, Ron Penny and Maxine Shephard, who read and commented on drafts of the manuscript, to Su Williams and Bob Willson, who assisted with computer programming, and to my Head of Department, Professor A. T. Welford for his encouragement. I am equally indebted to those responsible for the production of the final manuscript which was organised by Margaret Blaber ably assisted by Judy Hallett. My thanks also to Sue Thom who prepared the diagrams, and to my wife Kathie, who did the proof reading.

The impetus for this work came from a project on the applications of signal detection theory to the processing of verbal information, supported by Grant No A67/16714 from the Australian Research Grants Committee. I am also grateful to St John's College, Cambbridge, for making it possible to return to England during 1969 to work on the book, and to Adelaide University, which allowed me to take up the St John's offer.

A final word of thanks is due to some people who know more about the development of this book than anyone else. These are the Psychology III students at Adelaide University who have served as a tolerant but critical proving ground for the material which follows.

Adelaide University   D. MCNICOL
September 1970

Contents

Chapter 1

WHAT ARE STATISTICAL DECISIONS?

AN EXAMPLE

Often we must make decisions on the basis of evidence which is less than perfect. For instance, a group of people has heights ranging from 5 ft 3 in. to 5 ft 9 in. These heights are measured with the group members standing in bare feet. When each person wears shoes his height is increased by 1 inch, so that the range of heights for the group becomes 5 ft 4 in. to 5 ft 10 in. The distributions of heights for members of the group with shoes on and with shoes off are illustrated in the histograms of Figure 1.1.

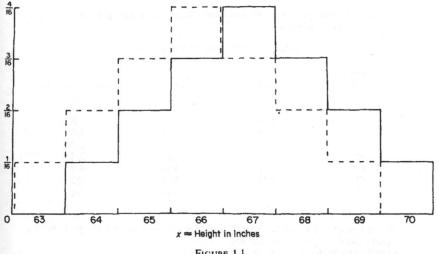

Solid line: Distribution s–shoes on
Dotted line: Distribution n–shoes off

x = Height in Inches

FIGURE 1.1

1

You can see that the two histograms are identical, with the exception that s, the 'Shoes on' histogram, is 1 in. further up the X-axis than n, the 'Shoes off' histogram.

Given these two distributions you are told that a particular person is 5 ft 7 in. tall and from this evidence you must deduce whether the measurement was taken with shoes on or with shoes off. A look at these histograms in Figure 1.1 shows that you will not be able to make a decision which is certain to be correct. The histograms reveal that 3/16ths of the group is 5 ft 7 in. tall with shoes off and that 4/16ths of the group is 5 ft 7 in. tall with shoes on. The best bet would be to say that the subject had his shoes on when the measurement was taken. Furthermore, we can calculate the odds that this decision is correct. They will be (4/16)/(3/16), that is, 4/3 in favour of the subject having his shoes on.

You can see that with the evidence you have been given it is not possible to make a completely confident decision one way or the other. The best decision possible is a statistical one based on the odds favouring the two possibilities, and that decision will only guarantee you being correct four out of every seven choices, on the average.

It is possible to calculate the odds that each of the eight heights

TABLE 1.1   *The odds favouring the hypothesis 'Shoes on' for the eight possible heights of group members.*

Height in inches x	Probability of obtaining this height with		Odds favouring s l(x)
	Shoes off (n) $P(x \mid n)$	Shoes on (s) $P(x \mid s)$	
63	1/16	0	0
64	2/16	1/16	1/2
65	3/16	2/16	2/3
66	4/16	3/16	3/4
67	3/16	4/16	4/3
68	2/16	3/16	3/2
69	1/16	2/16	2/1
70	0	1/16	

$P(x \mid n)$ and $P(x \mid s)$ are called 'conditional probabilities' and are the probabilities of x given n, and of x given s, respectively.

$l(x)$ is the symbol for the 'odds' or likelihood ratio.

2

of the group was obtained with shoes on. This is done in Table 1.1. The probabilities in columns 2 and 3 have been obtained from Figure 1.1.

For the sake of brevity we will refer to the two states of affairs 'Shoes on' and 'Shoes off' as states s and n respectively.

It can be seen that the odds favouring hypothesis s are calculated in the following way:

For a particular height, which we will call x, we take the probability that it will occur with shoes on and divide it by the probability that it will occur with shoes off. We could, had we wished, have calculated the odds favouring hypothesis n rather than those favouring s, as has been done in Table 1.1. To do this we would have divided column 2 entries by column 3 entries and the values in column 4 would then have been the reciprocals of those which appear in the table.

Looking at the entries in column 4 you will see that as the value of x increases the odds that hypothesis s is correct become more favourable. For heights of 67 in. and above it is more likely that hypothesis s is correct. Below x = 67 in. hypothesis n is more likely to be correct. If you look at Figure 1.1 you will see that from 67 in. up, the histogram for 'Shoes on' lies above the histogram for 'Shoes off'. Below 67 in. the 'Shoes off' histogram is higher.

SOME DEFINITIONS

With the above example in mind we will now introduce some of the terms and symbols used in signal detection theory.

The evidence variable

In the example there were two relevant things that could happen. These were state s (the subject had his shoes on) and state n (the subject had his shoes off). To decide which of these had occurred, the observer was given some evidence in the form of the height, x, of the subject. The task of the observer was to decide whether the evidence favoured hypothesis s or hypothesis n.

As you can see we denote evidence by the symbol x.[1] Thus x is called the *evidence variable*. In the example the values of x ranged

[1] Another symbol used by Green & Swets (1966) for evidence is e.

3

from $x = 63$ in. to $x = 70$ in. In a psychological experiment x can be identified with the sensory effect produced by a stimulus which may be, for example, a range of illumination levels, sound intensities, or verbal material of different kinds.

Conditional probabilities

In the example, given a particular value of the evidence variable, say $x = 66$ in., Table 1.1 can be used to calculate two probabilities:

(a) $P(x|s)$: that is, the probability that the evidence variable will take the value x given that state s has occurred. In terms of the example, $P(x|s)$ is the probability that a subject is 66 in. tall given that he is wearing shoes. From Table 1.1 it can be seen that for $x = 66$ in., $P(x|s) = \frac{3}{16}$.

(b) $P(x|n)$: the probability that the evidence variable will take the value x given that state n has occurred. Table 1.1 shows that for $x = 66$ in., $P(x|n) = \frac{4}{16}$.

$P(x|s)$ and $P(x|n)$ are called *conditional probabilities* because they represent the probability of one event occurring conditional on another event having occurred. In this case we have been looking at the probability of a person being 66 in. tall given that he is (or conditional on him) wearing shoes.

The likelihood ratio

It was suggested that one way of deciding whether state s or state n had occurred was to first calculate the odds favouring s. In signal detection theory, instead of speaking of 'odds' we use the term *likelihood ratio*. 'Odds' and 'likelihood ratio' are synonymous The likelihood ratio is represented symbolically as $l(x)$.

From the foregoing discussion it can be seen that in this example the likelihood ratio is obtained from the formula[1]

---

[1] More correctly we should write $l_{sn}(x_i) = P(x_i|s)/P(x_i|n)$, with the subscripts i, s and n, added to (1.1). The subscript i denotes the likelihood ratio for the ith value of x but normally we will just write x with the subscripts implied. The order of the subscripts s and n tell us which of $P(x_i|s)$ and $P(x_i|n)$ is to act as the denominator and numerator in the expression for the likelihood ratio. The likelihood ratio $l_{sn}(x_i)$ is the ratio of $P(x_i|s)$ to $P(x_i|n)$ where $P(x_i|s)$ serves as the numerator. On the other hand the likelihood ratio $l_{ns}(x_i)$ is the ratio of $P(x_i|n)$ to $P(x_i|s)$ where $P(x_i|n)$ serves as the numerator. As all likelihood ratios in this book will use probabilities involving s as the numerator and probabilities involving n as the denominator the s and n subscripts will be omitted.

$$lx = \frac{P(x \mid s)}{P(x \mid n)}. \tag{1.1}$$

Thus from Table 1.1 we can see that

$$l(x = 64) = \frac{1/16}{2/16}, \quad l(x = 66) = \frac{3/16}{4/16}, \quad \text{etc.}$$

Hits, misses, false alarms and correct rejections

We now come to four conditional probabilities which will be often referred to in the following chapters. They will be defined by referring to Table 1.1.

First, however, let us adopt a convenient convention for denoting the observer's decision.

The two possible stimulus events have been called s and n. Corresponding to them are two possible responses that an observer might make; observer says 's occurred' and observer says 'n occurred'. As we use the lower case letters s and n to refer to stimulus events, we will use the upper case letters S and N to designate the corresponding response events. There are thus four combinations of stimulus and response events. These along with their accompanying conditional probabilities are shown in Table 1.2.

TABLE 1.2 *The conditional probabilities, and their names, which correspond to the four possible combinations of stimulus and response events. The data in the table are the probabilities for the decision rule: 'Respond S if $x >$ 66 in.; respond N if $x \leqslant 66$ in.'*

		Response event		Row sum
		S	N	
Stimulus event	s	'Hit' $P(S \mid s) =$ $(4+3+2+1)/16$	'Miss' $P(N \mid s) =$ $(3+2+1)/16$	1·0
	n	'False alarm' $P(S \mid n) =$ $(3+2+1)/16$	'Correct rejection' $P(N \mid n) =$ $(4+3+2+1)/16$	1·0

The meanings of the conditional probabilities are best explained by referring to an example from Table 1.1. An observer decides to respond S when $x > 66$ in. and N when $x < 66$ in. The probability that he will say S given that s occurred can be calculated from column 3 of the table by summing all the $P(x|s)$ values which fall in or above the category $x = 66$ in., namely, $(4+3+2+1)/16 = 10/16$. This is the value of $P(S|s)$, the *hit rate* or *hit probability*. Also from column 3 we see that $P(N|s)$, the probability of responding N when s occurred is $(3+2+1)/16 = 6/16$. From column 2 $P(N|n)$, the probability of responding N when n occurred, is $10/16$, and $P(S|n)$, the *false alarm rate*, is $6/16$. These hits, misses, false alarms and correct rejections are shown in Table 1.2.

DECISION RULES AND THE CRITERION

The meaning of β

In discussing the example it has been implied that the observer should respond N if the value of the evidence variable is less than or equal to 66 in. If the height is greater than or equal to 67 in. he should respond S. This is the observer's *decision rule* and we can state it in terms of likelihood ratios in the following manner:

'If $l(x) < 1$, respond N; if $l(x) \geqslant 1$, respond S.'

Check Table 1.1 to convince yourself that stating the decision rule in terms of likelihood ratios is equivalent to stating it in terms of the values of the evidence variable above and below which the observer will respond S or N.

Another way of stating the decision rule is to say that the observer has set his *criterion* at $\beta = 1$. In essence this means that the observer chooses a particular value of $l(x)$ as his criterion. Any value falling below this criterion value of $l(x)$ is called N, while any value of $l(x)$ equal to or greater than the criterion value is called S. This criterion value of the likelihood ratio is designated by the symbol β.

Two questions can now be asked. First, what does setting the criterion at $\beta = 1$ achieve for the observer? Second, are there other decision rules that the observer might have used?

Maximizing the number of correct responses

If, in the example, the observer chooses the decision rule: 'Set the criterion at $\beta = 1$ in., he will make the maximum number of correct responses for those distributions of s and n. This can be checked from Table 1.1 as follows:

If he says N when $l(x) < 1$ he will be correct 10 times out of 16, and incorrect 6 times out of 16. If he says S when $l(x) \geq 1$ he will be correct 10 times out of 16, and incorrect 6 times out of 16. Overall, his chances of making a correct response will be 20/32 and his chances of making an incorrect response will be 12/32.

Can the observer do better than this? Convince yourself that he cannot by selecting other decision rules and using Table 1.1 to calculate the proportion of correct responses. For example, if the observer adopts the rule: 'Say N if $l(x) < \frac{3}{4}$ and say S if $l(x) \geq \frac{3}{4}$,' his chances of making a correct decision will be 19/32, less than those he would have had with $\beta = 1$.

It is a mistake, however, to think that setting the criterion at $\beta = 1$ will always maximize the number of correct decisions. This will only occur in the special case where an event of type s has the same probability of occurrence as an event of type n, or, to put it in symbolic form, when $P(s) = P(n)$. In our example, and in many psychological experiments, this is the case.

When s and n have different probabilities of occurrence the value of β which will maximize correct decisions can be found from the formula

$$\beta = P(n)/P(s) \tag{1.2}$$

We can see how this rule works in practice by referring to the example in Table 1.1.

Assume that in the example $P(s) = \frac{1}{2} P(n)$. Therefore by formula (1.2) $\beta = 2$ will be the criterion value of $l(x)$ which will maximize correct responses. This criterion is twice as strict as the one which

TABLE 1.3 *The number of correct and incorrect responses for $\beta = 1$ when $P(s) = \frac{1}{2} P(n)$.*

		Observer's response		Total (out of 48)
		S	N	
Stimulus event	s	10	6	16
	n	6×2	10×2	32
				48

Number of correct responses (out of 48) $= 10 + (10 \times 2) = 30$
Number of incorrect responses (out of 48) $= 6 + (6 \times 2) = 18$

7

B

maximized correct responses for equal probabilities of s and n. First we can calculate the proportion of correct responses which would be obtained if the criterion were maintained at $\beta = 1$. This is done in Table 1.3. As n events are twice as likely as s events, we multiply entries in row n of the table by 2.

The same thing can be done for $\beta = 2$. Table 1.1 shows that $\beta = 2$ falls in the interval $x = 69$ in. so the observer's decision rule will be: 'Respond S if $x \geqslant 69$ in., respond N if $x < 69$ in. Again, with the aid of Table 1.1, the proportion of correct and incorrect responses can be calculated. This is done in Table 1.4.

TABLE 1.4 *The number of correct and incorrect responses for $\beta = 2$ when $P(s) = \frac{1}{2} P(n)$.*

		Observer's response		
		S	N	Total (out of 48)
Stimulus event	s	3	13	16
	n	1×2	15×2	32
				48

Number of correct responses (out of 48) $= 3 + (15 \times 2) = 33$
Number of incorrect responses (out of 48) $= 13 + (1 \times 2) = 15$

It can be seen that $\beta = 2$ gives a higher proportion of correct responses than $\beta = 1$ when $P(s) = \frac{1}{2} P(n)$. There is no other value of β which will give a better result than 33/48 correct responses for these distributions of s and n.

Other decision rules

One or two other decision rules which might be used by observers will now be pointed out. A reader who wishes to see these discussed in more detail should consult Green & Swets (1966) pp. 20–7. The main purpose here is to illustrate that there is no one correct value of $l(x)$ that an observer should adopt as his criterion. The value of β he should select will depend on the goal he has in mind and this goal may vary from situation to situation. For instance the observer may have either of the following aims.

(a) *Maximizing gains and minimizing losses*. Rewards and penalties may be attached to certain types of response so that

$$V_s S = \text{value of making a hit,}$$
$$C_s N = \text{cost of making a miss,}$$
$$C_n S = \text{cost of making a false alarm,}$$
$$V_n N = \text{value of making a correct rejection.}$$

In the case where $P(s) = P(n)$ the value of β which will maximize the observer's gains and minimize his losses is

$$\beta = \frac{V_{nN} + C_{nS}}{V_{sS} + C_{sN}}. \tag{1.3}$$

It is possible for a situation to occur where $P(s)$ and $P(n)$ are not equal and where different costs and rewards are attached to the four combinations of stimuli and responses. In such a case the value of the criterion which will give the greatest net gain can be calculated combining (1.2) with (1.3) so that

$$\beta = \frac{(V_{nN} + C_{nS}) \cdot P(n)}{(V_{sS} + C_{sN}) \cdot P(s)}. \tag{1.4}$$

It can be seen from (1.4) that if the costs of errors equal the values of correct responses, the formula reduces to (1.2). On the other hand, if the probability of s equals the probability of n, the formula reduces to (1.3).

(b) *Keeping false alarms at a minimum*: Under some circumstances an observer may wish to avoid making mistakes of a particular kind. One such circumstance with which you will already be familiar occurs in the conducting of statistical tests. The statistician has two hypotheses to consider; H_0 the null hypothesis, and H_1, the experimental hypothesis. His job is to decide which of these two to accept. The situation is quite like that of deciding between hypotheses n and s in the example we have been discussing.

In making his decision the statistician risks making one of two errors:

Type I error : accepting H_1 when H_0 was true, and
Type II error : accepting H_0 when H_1 was true.

9

The Type I errors are analogous to false alarms and the Type II errors are analogous to misses. The normal procedure in hypothesis testing is to keep the proportion of Type I errors below some acceptable maximum. Thus we set up confidence limits of, say, $p = 0·05$, or, in other words, we set a criterion so that $P(S \mid n)$ does not exceed 5%. As you should now realize, by making the criterion stricter, not only will false alarms become less likely but hits will also be decreased. In the language of hypothesis testing, Type I errors can be avoided only at the expense of increasing the likelihood of Type II errors.

SIGNAL DETECTION THEORY AND PSYCHOLOGY

The relevance of signal detection theory to psychology lies in the fact that it is a theory about the ways in which choices are made. A good deal of psychology, perhaps most of it, is concerned with the problems of choice. A learning experiment may require a rat to choose one of two arms of a maze or a human subject may have to select, from several nonsense-syllables, one which he has previously learned. Subjects are asked to choose, from a range of stimuli, the one which appears to be the largest, brightest or most pleasant. In attitude measurement people are asked to choose, from a number of statements, those with which they agree or disagree. References such as Egan & Clarke (1966), Green & Swets (1966) and Swets (1964) give many applications of signal detection theory to choice behaviour in a number of these areas.

Another interesting feature of signal detection theory, from a psychological point of view, is that it is concerned with decisions based on evidence which does not unequivocally support one out of a number of hypotheses. More often than not, real-life decisions have to be made on the weight of the evidence and with some uncertainty, rather than on information which clearly supports one line of action to the exclusion of all others. And, as will be seen, the sensory evidence on which perceptual decisions are made can be equivocal too. Consequently some psychologists have found signal detection theory to be a useful conceptual model when trying to understand psychological processes. For example, John (1967) has proposed a theory of simple reaction times based on signal detection theory;

10

Welford (1968) suggests the extension of detection theory to absolute judgement tasks where a subject is required to judge the magnitude of stimuli lying on a single dimension; Boneau & Cole (1967) have developed a model for decision-making in lower organisms and applied it to colour discrimination in pigeons; Suboski (1967) has applied detection theory in a model of classical discrimination conditioning.

The most immediate practical benefit of the theory, however, is that it provides a number of useful measures of performance in decision-making situations. It is with these that this book is concerned. Essentially the measures allow us to separate two aspects of an observer's decision. The first of these is called *sensitivity*, that is, how well the observer is able to make correct judgements and avoid incorrect ones. The second of these is called *bias*, that is, the extent to which the observer favours one hypothesis over another independent of the evidence he has been given. In the past these two aspects of performance have often been confounded and this has lead to mistakes in interpreting behaviour.

Signal and noise

In an auditory detection task such as that described by Egan, Schulman & Greenberg (1959) an observer may be asked to identify the presence or absence of a weak pure tone embedded in a burst of *white noise*. (Noise, a hissing sound, consists of a wide band of frequencies of vibration whose intensities fluctuate randomly from moment to moment. An everyday example of noise is the static heard on a bad telephone line, which makes speech so difficult to understand.) On some trials in the experiment the observer is presented with noise alone. On other trials he hears a mixture of tone + noise. We can use the already familiar symbols s and n to refer to these two stimulus events. The symbol n thus designates the event 'noise alone' and the symbol s designates the event 'signal (in this case the tone) + noise'.

The selection of the appropriate response, S or N, by the observer raises the same problem of deciding whether a subject's height had been measured with shoes on or off. As the noise background is continually fluctuating, some noise events are likely to be mistaken for signal + noise events, and some signal + noise events will appear

11

to be like noise alone. On any given trial the observer's best decision will again have to be a statistical one based on what he considers are the odds that the sensory evidence favours *s* or *n*.

Visual detection tasks of a similar kind can also be conceived. The task of detecting the presence or absence of a weak flash of light against a background whose level of illumination fluctuates randomly is one which would require observers to make decisions on the basis of imperfect evidence.

Nor is it necessary to think of noise only in the restricted sense of being a genuinely random component to which a signal may or may not be added. From a psychological point of view, noise might be any stimulus not designated as a signal, but which may be confused with it. For example, we may be interested in studying an observer's ability to recognize letters of the alphabet which have been presented briefly in a visual display. The observer may have been told that the signals he is to detect are occurrences of the letter 'X' but that sometimes the letters 'K'. 'Y' and 'N' will appear instead. These three non-signal letters are not noise in the strictly statistical sense in which white noise is defined, but they are capable of being confused with the signal letter, and, psychologically speaking, can be considered as noise.

Another example of this extended definition of noise may occur in the context of a memory experiment. A subject may be presented with the digit sequence '58932' and at some later time he is asked: 'Did a "9" occur in the sequence?', or, alternatively: 'Did a "4" occur in the sequence?' In this experiment five digits out of a possible ten were presented to be remembered and there were five digits not presented. Thus we can think of the numbers 2, 3, 5, 8, and 9, as being signals and the numbers 1, 4, 6, 7, and 0, as being noise. (See Murdock (1968) for an example of this type of experiment.)

These two illustrations are examples of a phenomenon which, unfortunately, is very familiar to us—the fallibility of human perception and memory. Sometimes we 'see' the wrong thing or, in the extreme case of hallucinations, 'see' things that are not present at all. False alarms are not an unusual perceptual occurrence. We 'hear' our name spoken when in fact it was not; a telephone can appear to ring if we are expecting an important call; mothers are prone to 'hear' their babies crying when they are peacefully asleep.

12

Perceptual errors may occur because of the poor quality or ambiguity of the stimulus presented to an observer. The letter 'M' may be badly written so that it closely resembles an 'N'. The word 'bat', spoken over a bad telephone line, may be masked to such an extent by static that it is indistinguishable from the word 'pat'. But this is not the entire explanation of the perceptual mistakes we commit. Not only can the stimulus be noisy but noise can occur within the perceptual system itself. It is known that neurons in the central nervous system can fire spontaneously without external stimulation. The twinkling spots of light seen when sitting in a dark room are the result of spontaneously firing retinal cells and, in general, the continuous activity of the brain provides a noisy background from which the genuine effects of external signals must be discriminated (Pinneo, 1966). FitzHugh (1957) has measured noise in the ganglion cells of cats, and also the effects of a signal which was a brief flash of light of near-threshold intensity. The effects of this internal noise can be seen even more clearly in older people where degeneration of nerve cells has resulted in a relatively higher level of random neural activity which results in a corresponding impairment of some perceptual functions (Welford, 1958). Another example of internal noise of a rather different kind may be found in schizophrenic patients whose cognitive processes mask and distort information from the outside world causing failures of perception or even hallucinations.

The concept of internal noise carries with it the implication that all our choices are based on evidence which is to some extent unreliable (or noisy). Decisions in the face of uncertainty are therefore the rule rather the exception in human choice behaviour. An experimenter must expect his subjects to 'perceive' and 'remember' stimuli which did not occur (for the most extreme example of this see Goldiamond & Hawkins, 1958). So, false alarms are endemic to a noisy perceptual system, a point not appreciated by earlier psychophysicists who, in their attempts to measure thresholds, discouraged their subjects from such 'false perceptions'. Similarly, in the study of verbal behaviour, the employment of so-called 'corrections for chance guessing' was an attempt to remove the effects of false alarms from a subject's performance score as if responses of this type were somehow improper.

13

The fact is, if noise does play a role in human decision-making, false alarms are to be expected and should reveal as much about the decision process as do correct detections. The following chapters of this book will show that it is impossible to obtain good measures of sensitivity and bias without obtaining estimates of both the hit and false alarm rates of an observer.

A second consequence of accepting the importance of internal noise is that signal detection theory becomes something more than just another technique for the special problems of psychophysicists. All areas of psychology are concerned with the ways in which the internal states of an individual affect his interpretation of information from the world around him. Motivational states, past learning experiences, attitudes and pathological conditions may determine the efficiency with which a person processes information and may also predispose him towards one type of response rather than another. Thus the need for measures of sensitivity and response bias applies over a wide range of psychological problems.

Egan (1958) was first to extend the use of detection theory beyond questions mainly of interest to psychophysicists by applying it to the study of recognition memory. Subsequently it has been employed in the study of human vigilance (Broadbent & Gregory, 1963a, 1965; Mackworth & Taylor, 1963), attention (Broadbent & Gregory, 1963b; Moray & O'Brien, 1967) and short-term memory (Banks, 1970; Murdock, 1965; Lockhart & Murdock, 1970; Norman & Wickelgren, 1965; Wickelgren & Norman, 1966). The effects of familiarity on perception and memory have been investigated by detection theory methods by Allen & Garton (1968, 1969) Broadbent (1967) and Ingleby (1968). Price (1966) discusses the application of detection theory to personality, and Broadbent & Gregory (1967), Dandeliker & Dorfman (1969), Dorfman (1967) and Hardy & Legge (1968) have studied sensitivity and bias changes in perceptual defence experiments.

Nor has detection theory been restricted to the analysis of data from human observers. Suboski's (1967) analysis of discrimination conditioning in pigeons has already been mentioned, and Nevin (1965) and Rilling & McDiarmid (1965) have also studied discrimination in pigeons by detection theory methods. Rats have received similar attention from Hack (1963) and Nevin (1964).

14

Problems

The following experiment and its data are to be used for problems 1 to 6.

In a card-sorting task a subject is given a pack of 450 cards, each of which has had from 1 to 5 spots painted on it. The distribution of cards with different numbers of spots is as follows:

Number of spots on card	Number of cards in pack
1	50
2	100
3	150
4	100
5	50

Before giving the pack to the subject the experimenter paints an extra spot on 225 cards as follows:

Original number of spots on card	Number of cards in this group receiving an extra spot
1	25
2	50
3	75
4	50
5	25

The subject is then asked to sort the cards in the pack into two piles; one pile containing cards to which an extra spot has been added and the other pile, of cards without the extra spot.

1. What is the maximum proportion of cards which can be sorted correctly into their appropriate piles?

2. State, in terms of x, the evidence variable, the decision rule which will achieve this aim.

3. If the subject stands to gain 1¢ for correctly identifying each card with an extra spot and to lose 2¢ for incorrectly classifying a

15

card as containing an extra spot, find firstly in terms of β, and secondly in terms of x, the decision rule which will maximize his gains and minimize his losses.

4. What proportions of hits and false alarms will the observer achieve if he adopts the decision rule $\beta = \frac{3}{2}$?

5. What will $P(N \mid s)$ and β be if the subject decides not to allow the false alarm probability to exceed $\frac{2}{3}$?

6. If the experimenter changes the pack so that there are two cards in each group with an extra spot to every one without, state the decision rule both in terms of x and in terms of β which will maximize the proportion of correct responses.

7. Find the likelihood ratio for each value of x for the following data:

x	1	2	3	4
$P(x \mid n)$	0·2	0·4	0·5	0·6
$P(x \mid s)$	0·5	0·7	0·8	0·9

8. At a particular value of x, $l(x) = 0·5$ and the probability of x given that n has occurred is $0·3$. What is the probability of x given that s has occurred?

9. If $P(S \mid s) = 0·7$ and $P(N \mid n) = 0·4$, what is $P(N \mid s)$ and $P(S \mid n)$?

10. The following table shows $P(x \mid n)$ and $P(x \mid s)$ for a range of values of x.

x	63	64	65	66	67	68	69	70	71
$P(x \mid n)$	1/16	2/16	3/16	4/16	3/16	2/16	1/16	0	0
$P(x \mid s)$	1/16	1/16	2/16	2/16	4/16	2/16	2/16	1/16	1/16

Draw histograms for the distributions of signal and noise and compare your diagram with Figure 1.1. What differences can you see?

Find $l(x)$ for each x value in the table. Plot $l(x)$ against x for your data and compare it with a plot of $l(x)$ against x for the data in Table 1.1. How do the two plots differ?

If $P(s)$ were equal to $0·6$ and $P(n)$ to $0·4$ state, in terms of x, the decision rule which would maximize correct responses:

(a) for the problem data,
(b) for the data in Table 1.1.
(The issues raised in this problem will be discussed in Chapter 4.)

Chapter 2

NON-PARAMETRIC MEASURES OF SENSITIVITY

In Chapter 1 it was said that signal detection theory could be used both as the basis for a general theory of behaviour and for devising methods for measuring performance in experimental tasks. For the time being we will concern ourselves with this latter use and proceed to describing the basic types of detection task. There are three main types of experimental situation in which detection data can be collected; the yes–no task, the rating scale task and the forced-choice task.

THE YES–NO TASK
(Green & Swets, 1966, 32, ff.; Swets, Tanner & Birdsall, 1961)

The following is an example of a yes–no task. An observer watches a television screen on which, at regular intervals, some information appears. This information takes one of two forms. On half the occasions noise alone is shown. On other occasions noise plus a weak signal (a circular patch of light in the centre of the screen) is shown. Noise, and signal + noise trials occur at random. After each presentation the observer must say whether it was a signal + noise trial or just noise alone.

After a number of trials it is possible to construct a stimulus-response matrix summarising the results of the experiment. The form of the matrix is shown in Table 2.1. In it appear the four conditional probabilities, calculated from the subject's raw data, which were introduced in Table 1.2.

From the stimulus-response matrix we ought to be able to get some measure of how well the observer discriminates signal events

18

from those consisting of noise alone. At first sight it might be thought that $P(S|s)$, the hit rate, is a good index of the observer's sensitivity to signals. However it seems reasonable that some account should also be taken of $P(S|n)$, the false alarm rate. An observer who never looked at the display would be able to give a perfect hit rate by responding S to all trials. At the same time he would produce a false alarm rate equal to the hit rate and we would be unwilling to believe that such a perverse observer was showing any real sensitivity to signals.

TABLE 2.1 *The stimulus-response matrix for the yes–no task showing (a) the raw data obtained from 200 signal and 100 noise trials and (b) the conversion of the raw data into conditional probabilities by dividing each cell entry in the raw data table by its row total*

(a) raw data

		Response alternative		
		S	N	Row total
Stimulus event	s	150	50	200
	n	40	60	100

(b) Probabilities

		Response alternative		
		S	N	Row total
Stimulus event	s	$P(S\|s) = \dfrac{150}{200}$ $= 0.75$	$P(N\|s) = \dfrac{50}{200}$ $= 0.25$	1.00
	n	$P(S\|n) = \dfrac{40}{100}$ $= 0.40$	$P(N\|n) = \dfrac{60}{100}$ $= 0.60$	1.00

A way round the problem of obtaining a measure of the observer's sensitivity which takes into account both hits and false alarms is to begin by representing his results graphically. Although the stimulus-response matrix contains four probabilities we need only two of these to summarize his performance. If the hit rate is known, $P(N|s)$

19

can be obtained from $1 - P(S|s)$. If the false alarm rate is known then $P(N|n) = 1 - P(S|n)$. We therefore begin by plotting $P(S|s)$ as a function of $P(S|n)$. Suppose that in a yes–no task an observer gives a hit rate of 0·3 and a false alarm rate of 0·1. These values give us the single point d on curve A of Figure 2.1.

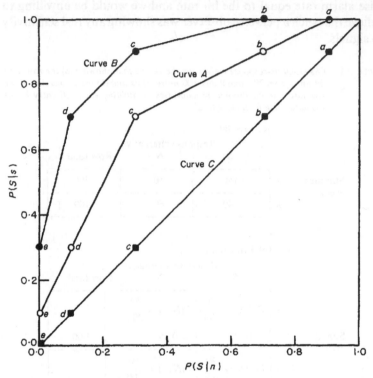

FIGURE 2.1   *Three* ROC *curves corresponding to the pairs of distributions, A, B, and C in Figure* 2.2

Keeping the same strengths of signal and noise we repeat the experiment, using the same observer, but encouraging him to be a little less cautious in his identification of signals. By doing this we have asked him to change his decision rule so that he accepts less certain evidence for making a S response. Note that nothing else has changed. Objectively speaking, signals are no more easy to

20

identify. The observer's sensitivity is the same as when he was using a stricter criterion for accepting a piece of sensory evidence as being a signal. What is to be expected then, is that with the less strict criterion the observer will give a higher hit rate but at the expense of a larger number of false alarms. With this new criterion suppose that the hit rate increases to 0·7 and the false alarm rate to 0·3. This will give us a second point (point c on curve A of Figure 2.1) which represents an equivalent level of sensitivity to the first point but a different degree of bias. The experiment can be repeated several times, on each occasion the observer being asked to use a stricter or a more lax criterion than on the previous occasion. By plotting $P(S|s)$ against $P(S|n)$ for each replication of the experiment we would trace the path of a curve along which sensitivity was constant but bias varied. Had the observer been asked to use five criteria of different degrees of strictness we may have ended up with the data in Table 2.2.

TABLE 2.2 *Hit and false alarm rates obtained from the same observer and with signal and noise strengths held constant, but with the observer varying his criterion for accepting evidence as signals*

	Observer's criterion	$P(S|s)$	$P(S|n)$
Very strict	e	0·1	0·0
	d	0·3	0·1
to	c	0·7	0·3
	b	0·9	0·7
Very lax	a	1·0	0·9

In Figure 2.1 the points corresponding to the $P(S|s)$ and $P(S|n)$ values of Table 2.2 are plotted out and joined together to give the curve labelled A. In this curve two other points have been included; $P(S|s) = P(S|n) = 0·0$, the strictest criterion the observer could ever have used when he is unwilling to accept any piece of evidence as signal, and $P(S|s) = P(S|n) = 1·0$, the laxest possible criterion when all events are accepted as signals. The curve obtained by joining all these points is called the *Receiver-Operating Characteristic* or ROC curve. It represents the various modes of observing behaviour (i.e. the various degrees of response bias) adopted when

21

the stimulus conditions (i.e. the strengths of signal and noise) are held constant. As will become apparent presently, this curve can be used to show how well the observer distinguishes *s* from *n* events.

Before this is done it should be noted that the data in Table 2.2 can be used to give some idea of the underlying distributions of signal and noise which gave rise to the observer's hit and false alarm rates.

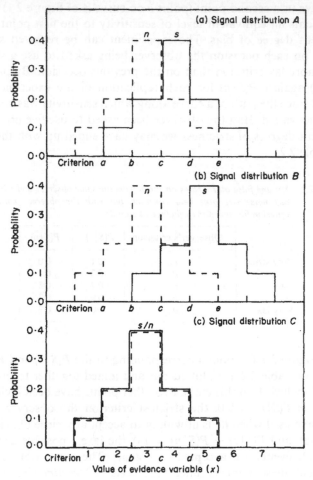

FIGURE 2.2   *Three pairs of signal and noise distributions with different distances between the distribution means*

In Figure 2.2(a) the x-axis shows the positions of the five criterion points, a to e, and the y-axis shows the values of $P(x|s)$ and $P(x|n)$. We get from Table 2.2 to Figure 2.2(a) by the following argument.

Starting with criterion e, Table 2.2 shows that at this point the hit rate is 0·1 and the false alarm rate is 0·0. This implies that 0·1 of the signal distribution must lie above e so if, in Figure 2.2(a), x has a maximum value of 6, the height of the signal distribution at this value of x which lies immediately above e must be 0·1. As the false alarm rate is 0·0 at e none of the noise distribution can lie above e. Moving back to d, a less strict criterion, we see that $P(S|s) = 0·3$ and $P(S|n) = 0·1$. Thus 0·3 of the signal distribution lies above d, and as we know that 0·1 of it is above e there must be $0·3 - 0·1 = 0·2$ between d and e. We can therefore add 0·2 to the signal histogram between d and e and 0·1 to the noise histogram. By continuing this process for points c, b and a a picture of the two distributions can be completed. Thus it is possible to work backwards from observed hit and false alarm rates to the distributions of signal and noise which gave rise to them.

Now look at Figures 2.2(b) and 2.2(c), both of which resemble Figure 2.2(a). Both have the same criterion points as Figure 2.2(a). The difference in Figure 2.2(b) is that the signal distribution has been moved further up the x-axis. This has the effect of giving higher hit rates for each criterion point but the false alarms will be the same as those in Table 2.2. Such a situation could arise if the experiment was repeated with the original observer but the signals were made stronger while the noise was kept at the same value. In Figure 2.2(c) the signal distribution has been moved down the x-axis so that it coincides with the noise distribution. Here sensitivity will be worse than in the other two cases and the hit rates will fall to the same values as the false alarm rates for each of the five criteria. If the hit and false alarm rates for the five criteria in Figure 2.2(b) are plotted out, the ROC curve labelled B in Figure 2.1 is obtained. Doing the same thing for Figure 2.2(c) the ROC curve labelled C in Figure 2.1 is the result. This last curve is actually a straight line connecting the points (0,0) and (1,1) in Figure 2.1. It can now be seen that the greater the separation between signal and noise distributions, the greater will be the area under the corresponding ROC curves in Figure 2.1. In case C where there is no separation between signal and noise distri-

23

butions, 0·5 of the total area in Figure 2.1 lies below curve C. For case A, a moderate degree of separation of the distributions, 0·73 of the total area lies below the ROC curve. In case B, where there is a high degree of separation, the ROC curve includes 0·89 of the total area. The best performance that an observer could give would be when 100 % of the total area in Figure 2.1 lies below the ROC curve.

A simple way of finding the area under the ROC curve is to plot the curve on graph paper and to count the number of squares beneath the curve. This number is then divided by the number of squares above the curve plus the number below (i.e. the total area) to give the proportion of the area beneath the curve. This proportion will be denoted by the symbol $P(A)$ (Green & Swets, 1966, 45-50, 404-5). While counting squares is an adequate method for the experimenter who has only a few ROC curves to deal with, the prospects of doing this for twenty or thirty curves is daunting. In Chapter 5 a geometrical method for finding areas will be described. This has the virtue of being easily written as a computer programme which is a considerable time-saver for experimenters with large amounts of data.

We have now obtained a solution to the problem raised earlier; how to obtain a measure of sensitivity which takes into account both the hit and false alarm probabilities. To do this we need to conduct a series of yes–no tasks holding the strengths of signal and noise constant and asking the observer to vary the strictness of his criterion from task to task. The resulting hit rates can then be plotted against their corresponding false alarm rates to determine the path of the ROC curve. By measuring the proportion of the total area which lies beneath the curve a value of $P(A)$ between 0·5 and 1·0 is obtained which is a direct index of the observer's ability to distinguish signal from noise events. As has been seen, the area under the curve is an index of the degree of overlap of the signal and noise distributions. A high degree of overlap means low sensitivity. Little overlap means high sensitivity.

The method proposed for obtaining the hit and false alarm rates may, however, be difficult for some types of experiment. Several hundred observations may be necessary to estimate each point on the curve with a reasonable degree of accuracy. The experimenter will have to guard carefully against factors such as practice and

24

fatigue which may alter the observer's sensitivity during the course of the experiment, for the method assumes that all points on the ROC curve represent equivalent degrees of sensitivity. There are a number of alternative techniques which offer various ways round the problem of the large number of trials required by the yes–no procedure. Some do not necessitate obtaining several hit and false alarm rates by which to trace the path of the curve. One, which will be discussed now, enables a number of hit and false alarm probabilities to be obtained quite efficiently and requires only a simple modification to the basic yes–no experiment.

THE RATING SCALE TASK
(Egan, Schulman & Greenberg, 1959; Green & Swets, 1966, 40–3; Pollack & Decker, 1958)

We have seen that an ROC curve can be constructed by inducing an observer to alter his criterion for accepting evidence as a signal and recording the hit and false alarm probabilities associated with each criterion level. In the basic yes–no task the observer is allowed only two responses S and N. However, he may be able to give us more information about his decision than this. In the rating scale task the observer is allowed several response categories. For example he may be permitted to classify each stimulus event into one of the four categories:

'Definitely a signal'
'Possibly a signal'
'Possibly noise'
'Definitely noise'

The rating scale allows responses to be graded on a scale of certainty that the evidence was a signal rather than forcing an all-or-none decision in the manner of the yes–no task. As many steps can be included on the rating scale as the experimeter desires. However, we have to be realistic about the number of confidence categories an observer can use consistently and we will not be far wrong if we use 4 to 10 steps on such a scale. There are exceptions to this rule of thumb and the practical details involved in the choice of a rating scale will be discussed more extensively in Chapter 5.

Notice the similarity between the rating scale procedure and the

25

basic yes–no task. Each point on the rating scale can be considered as corresponding to a different criterion. Points on the rating scale which indicate high certainty that a stimulus event was a signal correspond to strict criteria and we would expect these to give low false alarm rates. Points on the rating scale which indicate uncertainty as to whether an event was a signal or not will be laxer criteria and should give higher false alarm rates. It should be noted that being quite certain that an event was noise is equivalent to being quite uncertain that it was signal, so that for a three-point rating scale with the categories 'Certain signal', 'Uncertain either way', 'Certain noise', the strictest category will be the first and it should yield the lowest false alarm rate, and the laxest category will be the last, and it should yield the highest false alarm rate. How the hit and false alarm rates are calculated for each rating scale category will be seen presently.

The procedure of a rating scale experiment follows that of the simple yes–no task. In the observation interval the observer is presented with a stimulus which consists either of signal+noise or of noise alone. He then responds by indicating a point on the rating scale. After a series of trials (possibly still several hundred) the observer will have used all of the categories on the rating scale sufficiently often for an ROC curve to be constructed. In the practice trials preceding the experiment proper it is wise to encourage the observer to make full use of all the rating categories provided, as sometimes a particular observer may be initially reluctant to use either the categories of high confidence at the extreme ends of the scale or the middle categories. What to do when, despite the experimenter's best efforts, an observer remains recalcitrant about using some rating categories will be discussed in Chapter 5. In the meantime let us assume that the observer has been cooperative in this respect. After the experiment the observer's responses are sorted according to the confidence ratings they received and within each rating category we separate responses to signals from those to noise alone. The raw data for such an experiment may look like that in Table 2.3. This experiment uses a five-point confidence scale.

In this experiment signal and noise were presented 288 times each. It can be seen that the observer used the strictest signal (category 1) 172 times in all during the experiment and on these occasions he made 170 hits and 2 false alarms. The table of rating scale data looks like the table of data for the yes–no task (Table 2.1(a)). Both have two rows corresponding to the two stimulus events *s* and *n*. However in place of the two columns for the two responses *S* and *N* in the yes–no table, the rating scale table has five response categories.

TABLE 2.3   *Raw data from a rating scale task in which there were* 288 *signal and* 288 *noise trials and the observer gave his responses on a five-point confidence scale*

Observer's response
High certainty signal to low certainty signal

Category	1	2	3	4	5	Total
Stimulus *s*	170	52	21	25	20	288
n	2	11	17	67	191	288
Column Totals	172	63	38	92	211	576

How is the rating scale data used to plot an ROC curve? The first step is to convert the raw data in Table 2.3 into a set of hit and false alarm probabilities. To understand how this is done assume, for a moment, that instead of a rating scale task we had performed a simple yes–no task and that the criterion used by the observer was equivalent to the strictest of the five categories (i.e. category 1) used in the rating scale experiment. Restricted to only two responses, *S* or *N*, the observer would have produced the raw data of Table 2.4(a) which can be converted into the probabilities of Table 2.4(b).

It can be seen that the first column in Table 2.4(a), the *S* responses made by the observer, is the same as the first column in Table 2.3, the responses in category 1 of the rating scale. The second column in Table 2.4(a) is the same as the sums of the entries of columns 2, 3, 4 and 5 of Table 2.3. In effect, what we have done is to say that when the observer adopts the strictest category of the rating scale, all responses in that category are considered by him to be signals and all responses in the other four less strict categories are called noise. It is now an easy matter to calculate hit, false alarm, miss and correct rejection rates, as Table 2.4(b) shows.

27

TABLE 2.4   (a) *The matrix of raw scores which would have been obtained had the data from Table 2.3 been collected in a simple yes–no task with the observer using a criterion of equivalent strictness to category 1 of the rating scale.* (b) *The four conditional probabilities associated with category 1 of the rating scale*

(a) Raw data

| | | Response alternative | | |
		S	N	Row total
Stimulus event	s	170	52 + 21 + 25 + 20 = 118	288
	n	2	11 + 17 + 67 + 191 = 286	288

(b) Probabilities

| | | Response alternative | | |
		S	N	Row total
Stimulus event	s	$P(S\mid s) =$ 0·59	$P(N\mid s) =$ 0·41	1·00
	n	$P(S\mid n) =$ 0·01	$P(N\mid n) =$ 0·99	1·00

Next we wish to know the hit and false alarm rates associated with category 2 on the rating scale. As for category 1 we consider what results would have been obtained from a yes–no task using a single criterion of equivalent strictness to category 2. In such a task any response falling in category 2 or a stricter one would be called a signal by the observer and any response in a category less strict than category 2 would be called noise. Responses from categories 1 and 2 can thus be combined to give hits and false alarms, while collapsing categories 3, 4 and 5 into single totals for each row will give misses and correct rejections. This is shown in Table 2.5(a). This raw data can then be converted to probabilities, as Table 2.5(b) shows.

This process can be continued for the remaining categories of the rating scale. Notice when category 5 is reached that all the observer's responses will lie in this category or in a stricter one. Thus the hit rate and the false alarm rate for the last category of a rating scale will always be equal to 1·0 while the miss and correct

TABLE 2.5 *The matrix of raw scores which would have been obtained had the data from Table 2.3 been collected in a simple yes–no task with the observer using a criterion of equivalent strictness to category 2 of the rating scale. (b) The four conditional probabilities associated with category 2 of the rating scale*

(a) Raw data

Response alternative

		S	N	Row total
Stimulus event	s	170 + 52 = 222	21 + 25 + 20 = 66	288
	n	2 + 11 = 13	17 + 67 + 191 = 275	288

(b) Probabilities

Response alternative

		S	N	Row total
Stimulus event	s	$P(S\|s) =$ 0·77	$P(N\|s) =$ 0·23	1·00
	n	$P(S\|n) =$ 0·05	$P(N\|n) =$ 0·95	1·00

rejection rates will always be zero. As we know beforehand that the ROC curve will have to end at this point, the probabilities from the last category are not informative. It is worth bearing this in mind when designing a rating scale experiment. The number of useful points obtained from a rating scale task for the ROC curve will always be one less than the number of rating categories used.

The complete set of hit and false alarm rates for the raw rating data of Table 2.3 is shown in Table 2.6. They appear to behave sensibly. In the yes–no task it was seen that as less strict criteria were used

TABLE 2.6 *Hit and false alarm rates for each category from the rating scale data of Table 2.3*

Observer's response

High certainty signal to low certainty signal

Category	1	2	3	4	5
$P(S\|s)$	0·59	0·77	0·84	0·93	1·00
$P(S\|n)$	0·01	0·05	0·10	0·34	1·00

29

both hit and false alarm rates increased. In the rating scale task as less strict categories are used so also do hits and false alarms increase.

All that remains to be done is to plot the ROC curve for the hit and false alarm probabilities. This is done in Figure 2.3. The sensitivity measure $P(A)$, the proportion of the area beneath the ROC curve, can now be found and in this case turns out to be 0·92.

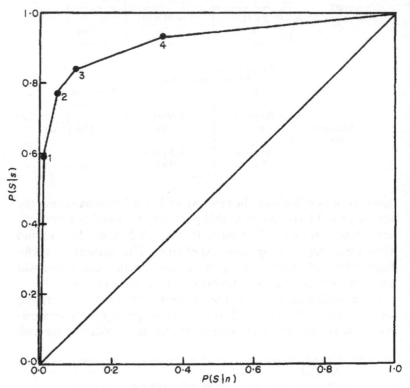

FIGURE 2.3   ROC *curves for the rating scale data in Table* 2.6

The rating scale method is more efficient than the yes–no task because all the data in Table 2.3 are used to determine each point on the ROC curve. In the yes–no task new data had to be collected

30

for each $P(S|s)$ and $P(S|n)$ value. Thus some 250 signal and 250 noise trials may be needed to find $P(S|s)$ and $P(S|n)$ for each criterion using yes–no procedure. With rating scale methods, 250 signal and 250 noise trials may suffice to find $P(S|s)$ and $P(S|n)$ for all the criteria. The basic difference between the two techniques is that in the yes–no task the observer adopts one criterion at a time. After $P(S|s)$ and $P(S|n)$ have been determined for the first criterion he is asked to adopt a different criterion and $P(S|s)$ and $P(S|n)$ are determined again, and so on, until the ROC curve can be plotted. In the rating scale task we capitalize on the fact that an observer can hold several criteria of different degrees of strictness simultaneously, thus allowing the determination of several pairs of hit and false alarm rates within the one block of trials.

Although rating scale procedure is more efficient than yes–no procedure there are penalties involved in using the technique. In the yes–no task where the observer was induced to change his criterion for each block of trials, the resulting points for the ROC curve were each based on a fresh set of observations and were therefore independent of each other (assuming that factors such as practice and fatigue have not affected the observer's performance). In the rating scale task all points for the ROC curve are based on the same set of observations and are thus heavily interdependent. When it comes to statistical analysis of results, the interdependence of the points obtained by the rating scale method can prove problematic. We may wish to test statistically whether two ROC curves represent different levels of sensitivity and, until recently, there was no way of doing this for curves based on rating scale data. Matters of statistical analysis will be mentioned in Chapter 5 but it is worth noting here that, despite its extra efficiency, the rating scale method is not always a satisfactory alternative to a series of yes–no tasks with criteria of different degrees of strictness.

AREA ESTIMATION WITH ONLY A SINGLE PAIR OF HIT AND FALSE ALARM RATES

It seems almost axiomatic that the area under an ROC curve can only be determined if we have a number of values of $P(S|s)$ and $P(S|n)$ from which to construct the curve. However, it has been shown by Norman (1964) and Pollack, Norman and Galanter (1964) that a single pair of $P(S|s)$ and $P(S|n)$ values provide enough information to determine approximately the path of the entire ROC curve.

Figure 2.4 shows the plot of a single pair of hit and false alarm rates for a point of i. The hit and false alarm probabilities for this

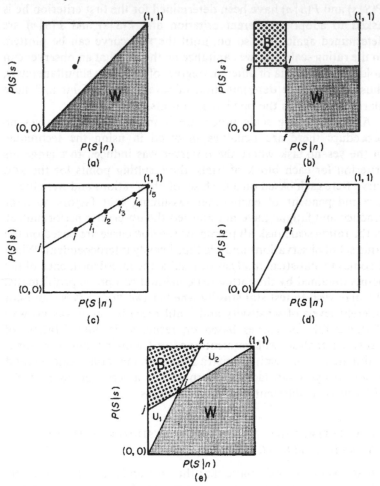

FIGURE 2.4 *Regions in which values of $P(S|s)$ and $P(S|n)$ will represent performance better or worse than that at point i*

32

point will be called $P_i(S|s)$ and $P_i(S|n)$. As the ROC curve on which i lies consists of a series of points each with equivalent sensitivity to i, the regions through which i's ROC curve cannot pass will be those where performance is either clearly better than or worse than that at i. Finding the path of i's ROC curve thus comes down to dividing the area of the graph into that which represents better performance than i, worse performance than i and equivalent performance to i.

Look first at Figure 2.4(a). You will remember that the straight line running from $(0, 0)$ to $(1, 1)$ is the ROC curve obtained from completely overlapping distributions of signal and noise; that is when the observer is completely unable to tell the difference between signal noise and $P(S|s) - P(S|n)$ for all points on the curve. All points in the shaded area below the line will give hit rates which are less than their corresponding false alarm rates and must therefore represent inferior performance to i whose hit rate is greater than its false alarm rate.

Next look at Figure 2.2(b). The area below the line from i to e and to the right of the line from i to f represents performance:

(a) when the observer makes more false alarms without improving hits, or

(b) makes fewer hits without making less false alarms.

This area must therefore represent worse performance than i does.

On the other hand, the area above the line from i to g and to the left of the line from i to h represents performance:

(c) when the observer makes fewer false alarms without making less hits, or

(d) makes more hits without increasing false alarms.

This area must therefore represent better performance than i does.

Thirdly, consider Figure 2.4(c). Suppose that the observer decides to alter his response bias in favour of making more S responses. We know from earlier discussion that this change of bias will increase $P(S|s)$ and $P(S|n)$ but, as the stimulus conditions have remained constant, sensitivity will not change. The observer changes his bias by increasing S responses by a proportion p. This means that a proportion of the responses which were misses will now become hits, and a proportion of the responses which were correct rejections

33

will become false alarms. These changes in the stimulus-response matrix are shown in Table 2.7.

You may be able to see that increasing $P(S|s)$ and $P(S|n)$ in this way results in the point i being moved along the straight line which joins i and $(1, 1)$ in Figure 2.4(c). If this is not obvious from the hit and false alarm values in Table 2.7 the following example may be helpful.

TABLE 2.7 *Changes in hit and false alarm rates if a bias to S responses results in $P(S|s)$ and $P(|n)$ being increased by a proportion p. Thus in (b) below $P(S|s)$ is increased by $p[1 - P(S|s)]$ while $P(N|s)$ is decreased by the same amount. $P(S|n)$ is increased by $p[1 - P(S|n)]$ while $P(N|n)$ is decreased by the same amount*

(a) The observer's performance at i

| | | Observer's response | | |
		S	N	Total
Stimulus event	s	$P_i(S|s)$	$P_i(N|s)$ $= 1 - P_i(S|s)$	1·00
	n	$P_i(S|n)$	$P_i(N|n)$ $= 1 - P_i(S|n)$	1·00

(b) The observer's performance if S responses are increased by a proportion p

| | | Observer's response | | |
		S	N	Total
Stimulus event	s	$P_i(S|s) + p[1 - P_i(S|s)]$	$1 - P_i(S|s) - p[1 - P_i(S|s)]$	1·00
	n	$P_i(S|n) + p[1 - P_i(S|n)]$	$1 - P_i(S|n) - p[1 - P_i(S|n)]$	1·00

The point i in Figure 2.4(c) represents $P(S|s) = 0\cdot60$ and $P(S|n) = 0\cdot30$. The observer now decides to increase S responses by a proportion $p = 0\cdot20$. The miss rate will thus be decreased by a factor of $0\cdot20$ and will become $0\cdot40 - (0\cdot20 \times 0\cdot40) = 0\cdot32$. This means that the hit rate will increase to the extent that the miss rate is decreased so it will become $0\cdot60 + (0\cdot20 \times 0\cdot40) = 0\cdot68$. Also, the correct rejection rate will be decreased by a factor of $0\cdot20$ and will drop from $0\cdot70$ to $0\cdot70 - (0\cdot20 \times 0\cdot70) = 0\cdot56$. At the same time false alarms will go up by the same amount as misses go down, so will increase from $0\cdot30$ to $0\cdot30 + (0\cdot20 \times 0\cdot70) = 0\cdot44$.

This is shown in Table 2.8. Row 1 of the table shows the proportion by which S responses are to be increased. The first entry in this row is for point i itself and as we are beginning at this point, the value of the proportion will be 0. The second entry in row 1 is the example we have just been considering where S responses are to be increased by a factor of 0·20. The second row of the table shows the extra

TABLE 2.8   *An example of the way in which $P(S|s)$ and $P(S|n)$ change if the observer alters his response bias by increasing S responses by a proportion p*

Point:	i	i_1	i_2	i_3	i_4	i_5
(1) p	0·00	0·20	0·40	0·60	0·80	1·00
(2) $p[1-P_i(S|s)]$	0·00	0·08	0·16	0·24	0·32	0·40
(3) $P_i(S|s)+p[1-P_i(S|s)]$	0·60	0·68	0·76	0·84	0·92	1·00
(4) $p[1-P_i(S|n)]$	0·00	0·14	0·28	0·42	0·56	0·70
(5) $P_i(S|n)+p[1-P_i(S|n)]$	0·30	0·44	0·58	0·72	0·86	1·00

proportion of hits which is to be added to the original hit rate at i. For $p = 0·20$ we had worked this out to be $0·20 \times 0·40 = 0·08$. Row 3 shows the hit rate for the new point. This is the hit rate at i (0·60) plus the extra proportion of hits ($0·60 + 0·08 = 0·68$). Row 4 gives the extra proportion of false alarms which is to be added to the original false alarm rate at i. For $p = 0·20$ this was $0·20 \times 0·70 = 0·14$ so that the new false alarm rate is the old false alarm rate at i (0·30) plus the extra proportion of false alarms (0·14) which equals 0·44. Row 5 gives the new false alarm rates.

To summarize Table 2.8:

(a) p is the proportion by which S responses are to be increased to give a new point of equivalent sensitivity to i.

(b) $p[1-P_i(S|s)]$ is the amount to be added to the hit rate for i to give the new hit rate for the point i_j.

(c) $P_i(S|s)+p[1-P_i(S|s)]$ is the hit rate for the new point i_j.

(d) $p[1-P_i(S|n)]$ is the amount to be added to the false alarm rate for i to give the new false alarm rate for point i_j.

(e) $P_i(S|n)+p[1-P_i(S|n)]$ is the false alarm rate for the new point i_j.

In the table, hit and false alarm rates have been calculated for five points. For each point, the hit rate has been plotted against

35

the false alarm rate in Figure 2.4(c) and it can be seen that these points fall on a straight line joining i with $(1, 1)$.

The same line can be extended down from i to the point j. This segment of the line would result from p being negative, that is if the observer was being more cautious and making less S responses than at i.

All the points on the line through j, i, and $(1, 1)$ might have been obtained simply by the observer changing his willingness to respond S and thus may represent equivalent levels of sensitivity.

Finally, consider Figure 2.4(d). Suppose now that the observer decides to alter his bias in favour of N responses. This will result in $P_i(S|s)$ and $P_i(S|n)$ being decreased. If the observer increases N responses by a proportion q, Table 2.9 shows the new hit and false alarm rates resulting from this bias change. The table is like Table 2.7 except that a proportion is subtracted from the hit rate and added to the miss rate, and a proportion is subtracted from the false alarm rate and added to the correct rejection rate.

TABLE 2.9   *Changes in hit and false alarm rates if a bias to N responses results in $P(S|s)$ and $P(S|n)$ being decreased by a proportion q. Thus in (b) above $P(S|s)$ is decreased by $qP(S|s)$ while $P(N|s)$ is increased by the same amount. $P(S|n)$ is decreased by $qP(S|n)$ while $P(N|n)$ is increased by the same amount*

(a) The observer's performance at i

		Observer's response	
		S	N
Stimulus event	s	$P_i(S\|s)$	$P_i(N\|s)$ $= 1 - P_i(S\|s)$	1·00
	n	$P_i(S\|n)$	$P_i(N\|n)$ $= 1 - P_i(S\|n)$	1·00

(b) The observer's performance if N responses are increased by a proportion q

		Observer's response		
		S	N	Total
Stimulus event	s	$P_i(S\|s) - qP_i(S\|s)$ $= (1-q)P_i(S\|s)$	$1 - P_i(S\|s) + qP_i(S\|s)$ $= 1 - (1-q)P_i(S\|s)$	1·00
	n	$P_i(S\|n) - qP_i(S\|n)$ $= (1-q)P_i(S\|n)$	$1 - P_i(S\|n) + qP_i(S\|n)$ $= 1 - (1-q)P_i(S\|n)$	1·00

Doing this results in point i being moved along a straight line which joins i with $(0, 0)$. If you wish to verify this select some values of q and find the hit and false alarm rates from the formulae in Table 2.9 using the same method as for the example in Table 2.8.

The line from $(0, 0)$ to i can be extended till it reaches the point k (see Figure 2.4(c). All points on this line may have been obtained from an observer changing his willingness to respond N and hence may represent equivalent levels of sensitivity.

In Figure 2.4(e), Figure 2.4(c) has been superimposed on Figure 2.4(d). In that figure the lines $(j, i, (1, 1))$ and $((0, 0), i, k)$ represent hit and false alarm rates of equivalent degrees of sensitivity to those at i. Any of the points on these lines may have been obtained purely by changes in response bias. Therefore any hit and false alarm rates which give points in the unshaded areas U_1 and U_2 might also have been obtained by a change in bias and need not represent greater sensitivity than i does. On the other hand, the shaded area B, which lies above the two lines, includes points, all of which represent better performance than i. (Notice that B includes the area of better performance in Figure 2.4b.) Also, the shaded area W which lies below the two lines must represent worse performance than i. (Notice that W includes the areas of worse performance in Figures 2.4a and 2.4b.)

As an ROC curve is the line on which hits and false alarms represent the same degree of sensitivity but different degrees of bias, we must conclude that the ROC curve on which i lies must pass through i and somewhere through the areas U_1 and U_2. It cannot pass through B and W.

Two useful things emerge from the procedure of approximately determining the ROC curve from a single pair of hit and false alarm rates.

First, if we have obtained two pairs of hit and false alarm rates for the same observer working under two experimental conditions we may wish to know whether sensitivity under one condition is superior to that under the other. To answer this question the hit and false alarm rates can be represented as a pair of points on a graph such as that of Figure 2.4. One point can then be selected and the lines passing from $(0, 0)$ through the point to the point k and passing from $(1, 1)$ through the point to the point j, can be drawn in. It can

37

then be seen if the other point lies in the area B (in which case it represents a higher sensitivity), W (in which case it represents a lower sensitivity), or one of the areas U_1 and U_2 (in which case no decision can be made). If a group of observers has yielded hit and false alarm rates under the two experimental conditions each observer's performance can be classified in this way and a sign test used to see whether the difference in sensitivity between the two conditions is statistically significant.

Second, it is possible, from a single pair of hit and false alarm rates, to obtain an approximate estimate of the area beneath the ROC curve and to use the approximation to $P(A)$ as a sensitivity score. From Figure 2.4 it can be seen that the greatest area which could lie beneath the ROC curve for i would be $W + U_1 + U_2$. The smallest area beneath the ROC curve for i would be equal to W. Therefore the maximum value of $P(A)$, the proportion of the area which lies below the ROC curve, is given by the formula

$$P(A) \quad (Upper \quad bound) = \frac{U_1 + U_2 + W}{U_1 + U_2 + W + B}. \tag{2.1}$$

The minimum value of $P(A)$ is given by the formula

$$P(A) \, (Lower \quad bound) = \frac{W}{U_1 + U_2 + W + B}. \tag{2.2}$$

Pollack & Norman (1964) propose that an estimate of $P(A)$ can be obtained by averaging (2.1) and (2.2). If we call this approximation to $P(A)$, $P(\bar{A})$ then:

$$P(\bar{A}) = \frac{\tfrac{1}{2}(U_1 + U_2) + W}{U_1 + U_2 + W + B}. \tag{2.3}$$

$P(\bar{A})$ is thus an estimate of sensitivity based on a single pair of hit and false alarm probabilities.

To save the tedium involved in finding $P(\bar{A})$ graphically from values of $P(S|s)$ and $P(S|n)$, tables of $P(\bar{A})$ have been prepared and appear in Appendix 5. By using these $P(\bar{A})$ can be read off directly from the appropriate values of $P(S|s)$ and $P(S|n)$.

A word of warning should be given at this point. $P(\bar{A})$ may be

38

a convenient measure of sensitivity but $P(A)$, calculated from several values of $P(S|s)$ and $P(S|n)$, is the more accurate estimate of the area under the ROC curve. Norman (1964) warns that comparing points by the use of this type of approximation involves the assumption that the ROC curves on which the points lie do not cross one another. Although no such curves have been encountered in this book to this point, Chapter 4 will show that crossing over can occur.

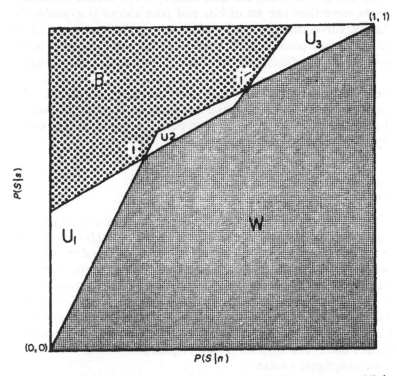

FIGURE 2.5   *The effect on the regions B, U, and W of having two points, i and i'. for the ROC curve*

Also it has been found by computer simulation that $P(\bar{A})$ gives the same value as $P(A)$ only if the observer is unbiased and does not have a tendency to give a larger proportion of S or N responses. The greater his response bias in either direction, the more $P(\bar{A})$ will under-estimate the true area under the ROC curve (problem 2 at the end of this chapter illustrates this effect). This means that $P(\bar{A})$ can only be used safely as a sensitivity estimate when response biases under the experimental conditions being examined are equivalent. If observers show greater bias in one condition than in another their sensitivity scores will appear to be lower, but this would be an artefact.

Finally, all that has been said about determining areas U, B and W for a single set of hits and false alarms can be applied to the case where more than one set of hits and false alarms is available for the ROC curve. Imagine that in addition to point i in Figure 2.4 we have a second point i' which represents the same degree of sensitivity as i but a different degree of bias. Figure 2.5 shows the regions U, B and W for the two points. Notice that the regions labelled U are decreased in size when more than one point is available. The more points we have, the better the determination of the path of the ROC curve. With three or four points well spaced out, the U region becomes very small indeed.

THE FORCED-CHOICE TASK
(Green & Swets, 1966, 43–5)

The forced-choice recognition task is commonly used in psychological experiments and psychological tests. It may take two forms. In the first the observer is presented with a stimulus and then shown two or more response alternatives. One of these is the stimulus; the remainder are incorrect. He must choose what he considers to be the correct response. In the second, commonly a psychophysical task, the observer is presented with two or more stimulus intervals. All but one of these contain noise; the remaining one contains signal + noise. The observer is then required to identify the interval containing signal + noise.

This second procedure can be used to turn the yes–no task described at the beginning of this chapter, into a forced-choice task. The observer sees two intervals, one containing the signal and the other, noise alone. In a series of trials, noise and signal + noise are allocated at random to the two intervals, and after each trial the observer attempts to identify the interval which contained signal + noise.

40

In this hypothetical example we will use, for the sake of comparison, the same distributions of signal and noise as were used in Figure 2.2(a) for the yes–no task. These distributions are shown again in Figure 2.6 as the distributions to be used in a two-alternative forced-choice task (or 2AFC task for short). As shown on the x-axis of the figure, x, the evidence variable, can take any of six values. In selecting stimuli for the two intervals of the task a value of x is chosen at random for the interval to contain noise alone according to the probabilities dictated by the noise histogram. Thus $x = 1$ will have a probability of 0·1 of selection for the noise interval, $x = 2$, a probability of selection of 0·2, etc. The other interval of each pair will contain a value of x selected at random from the six values of x for the signal distribution. Then, by random selection, it is decided whether the interval containing signal will come first or second in the trial. Thus, for the distributions in Figure 2.6 a series of forced-choice trials might be arranged like this:

Trial	Interval 1	Interval 2
1	Signal: $x = 3$	Noise: $x = 1$
2	Signal: $x = 6$	Noise: $x = 4$
3	Noise: $x = 3$	Signal: $x = 4$
4	Signal: $x = 4$	Noise: $x = 4$

and so on.

When a trial of two such intervals has been presented, how might the observer go about deciding which was the one containing the signal? As the signal distribution is further up the x-axis than the noise distribution, values of x drawn from the signal distribution will be, on the average, larger than those drawn from the noise distribution. A possible strategy, then, would be to observe which interval contained the larger x value, and to say that it was the one containing the signal.

Of course the observer will make some incorrect responses as some of the trials will have intervals in which the value of x for noise will be greater than the value of x for signal. However, as the average value of x for signal is greater than that for noise, the observer will be right more often than wrong.

Given the two distributions in Figure 2.6, what proportion of correct responses can the observer be expected to make? If he can

41

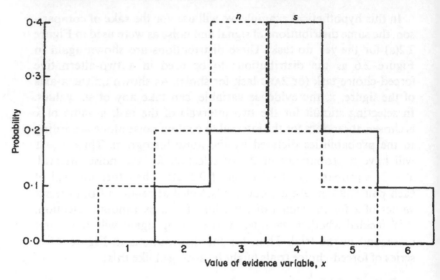

FIGURE 2.6 *Distributions of signal and noise for a 2-alternative forced-choice task*

reliably tell which of the two intervals contains the larger x, he will be correct as many times as the x in the signal interval exceeds the x in the noise interval. However, in this example there will be occasions when both noise and signal intervals have x's of the same value. In such a case the observer would have to make a random guess at which interval contained signal. In practice this state of affairs would rarely occur. For the sake of simplicity in the example x can take only one of six possible values. In a real experiment x usually varies continuously giving an infinite range of possible values and thus making it highly unlikely that both signal and noise would have the same x in the one trial.

To determine the proportion of correct responses expected in this 2AFC task we need to know the number of times that the x in the signal interval will be greater than the x in the noise interval. This is done in the following way:

The chances that the signal interval will contain an x equal to 2 can be found from the signal histogram to be 0.1. If $x = 2$ for signal, there is only one value of x for noise which is less than this, $x = 1$. The chances of the noise interval containing an x of 1 can be seen to be 0.1. So, the chances that on a particular trial of having $x = 2$ in the signal interval and $x = 1$ in the noise interval will be $0.1 \times 0.1 = 0.01$. Next the chances of having $x = 3$ in the signal interval can be seen to be equal to 0.2. The noise interval x will be less than the signal interval x when it is equal to either 2 or 1. The chances of getting signal $x = 3$ and noise $x = 2$ in the same trial will be $P(x = 3|s) \times P(x = 2|n) = 0.2 \times 0.2 = 0.04$. The chances of getting signal $x = 3$ and noise $x = 1$ in the same trial will be $P(x = 3|s) \times P(x = 1|n) = 0.2 \times 0.1 = 0.02$. We can now move to $x = 4$ for the signal interval where there are three cases, $= 1, 2$ or 3, when the noise x will be less than the signal x, find the three probabilities that this signal x will occur together with each of the noise x's and repeat the process for signal x's of 5 and 6. In each case we have been finding the *joint probability* of occurrence of signal x's with noise x's smaller than them. If we denote this joint probability by the symbols $P(x_s . x_n)$ which means the probability that a particular signal x will occur together with a particular noise x, then from the examples given above it is apparent that $P(x_s . x_n) = P(x|s) . P(x|n)$. $P(x|s)$ and $P(x|n)$ are, of course, the conditional probabilities introduced in Chapter 1 of obtaining a particular value of x given that either signal or noise occurred.

The calculation of all the joint probabilities is shown in Table 2.10. $P(x|s)$ is shown along the rows of the table, $P(x|n)$, along the columns, and $P(x_s . x_n)$ is the entry in each cell. The cells in the table enclosed below the solid line are those for joint occurrences of x_s and x_n where x_s is greater than x_n. For these occurrences the observer will choose the signal interval correctly. To find his overall chances of success we need to add up all the $P(x_s . x_n)$ values for these pairs of x_s and x_n. This total comes out to be 0.63. Now look at the cells along the diagonal of the table. These are enclosed within a dotted line. They represent joint occurrences of signal and noise x's when $x_s = x_n$. We have assumed that on these occasions the observer guesses randomly which interval contains the signal. As signal has equal chances of appearing in either interval, he has one chance in two of being correct. The sum of the $P(x_s . x_n)$s when $x_s = x_n$ is the sum of the diagonal entries which is 0.20. Therefore, to find the total proportion of correct responses in the 2AFC task we need the sum of the $P(x_s . x_n)$s when $x_s > x_n$ plus half the sum of the $P(x_s . x_n)$s

43

when $x_s = x_n$. This is $0.63 + (\frac{1}{2} \times 0.20) = 0.73$. Thus the observer should identify the signal correctly 0.73 of the time. This is the sensitivity score for the forced-choice task. It is called $P(c)$, the proportion of correct responses.

At this point let us refer again to ROC curve A in Figure 2.1. This curve was the result of a yes–no task based on the same signal and noise distributions as used in the 2AFC example. As the distributions were the same in both cases, it would be reasonable to expect that the observer's sensitivity would be the same whether he was tested

TABLE 2.10   *Determining $P(c)$ for the 2AFC task based on the distributions of signal and noise in Figure 2.6 from a matrix of joint probabilities of occurrence of each x_s and x_n value*

		x_n: 1	2	3	4	5	6
		$P(x\mid n)$: 0·1	0·2	0·4	0·2	0·1	0·0
x_s	$P(x\mid s)$						
1	0·0	0·00	0·00	0·00	0·00	0·00	0·00
2	0·1	0·01	0·02	0·04	0·02	0·01	0·00
3	0·2	0·02	0·04	0·08	0·04	0·02	0·00
4	0·4	0·04	0·08	0·16	0·08	0·04	0·00
5	0·2	0·02	0·04	0·08	0·04	0·02	0·00
6	0·1	0·01	0·02	0·04	0·02	0·01	0·00

Total sum of $P(x_s . x_n)$s $= 1.00$

Sum of $P(x_s . x_n)$s when $x_s > x_n = 0.63$
$\frac{1}{2}$ sum of $P(x_s . x_n)$s when $x_s = x_n = 0.10$
$P(c) = 0.63 + 0.10 = 0.73$.

by yes–no or forced-choice methods. In fact, $P(A)$, the area beneath the yes–no ROC curve was equal to 0.73 which is the same as $P(c)$ for the 2AFC task. Thus the 2AFC task is equivalent to the yes–no task. Given the $P(A)$ measure from a yes–no ROC curve we can predict the value of $P(c)$ for a 2AFC task and vice versa.

So far only the 2AFC task has been considered. Some experiments use more than a single incorrect alternative or more than one interval containing noise alone. When an observer must select the signal interval from m intervals, $m - 1$ of which contain noise alone, we would expect him to behave in much the same way as he does in a 2AFC task, inspecting all the intervals and choosing as the signal interval the one with the largest x.

$P(c)$ can be found for the mAFC task but it is commonly observed that as the number of intervals containing noise alone is increased, the value of $P(c)$ decreases. This happens despite the fact that signal and noise strengths have been held constant. That is to say, with sensitivity held constant, $P(c)$ varies inversely as the number of incorrect alternatives in the forced-choice task. Thus $P(c)$ for a 4AFC task will generally be less than $P(A)$, the area under the yes–no ROC curve. The reasons for this decline in $P(c)$ as the number of alternatives increases will be discussed in the next chapter. It should be noted here that while $P(c)$ from a 2AFC task can often be compared directly with $P(A)$, the area under the equivalent yes–no ROC curve, the correspondence between $P(c)$ from a mAFC task and $P(A)$, and between $P(c)_{2AFC}$ and $P(c)_{mAFC}$, is not so simple. The problem is discussed in detail by Green & Swets (1966, 50–51).

AN OVERALL VIEW OF NON-PARAMETRIC
SENSITIVITY MEASURES

So far we have derived three measures of sensitivity:

(a) $P(A)$, the proportion of the area beneath an ROC curve obtained either from a series of yes–no tasks where signals and noise strengths are held constant and bias varies, or from a single rating scale task where the observer holds several criteria simultaneously.

(b) $P(\bar{A})$, the approximate area under an ROC curve whose path has been estimated from a single pair of hit and false alarm probabilities.

(c) $P(c)$, the proportion of correct responses in a forced-choice task which, when a 2AFC task is used, equals $P(A)$ for an equivalent yes–no task.

Within limits these three measures will give the same sensitivity score. The potential experimenter may then be asking himself which, if any, is the best one to use. The answer to this question depends in part on the possibilities for collecting data that the particular experiment allows. In some cases a choice between yes–no

45

and rating scale methods will be dictated by the number of trials it is possible to conduct for each observer. In other cases the experiment may preclude the possibility of obtaining either by yes–no or by rating scale methods, a set of readings for different criteria.

All things being equal however the following points can be made. $P(A)$ is certainly a better measure of sensitivity than $P(\bar{A})$. With a number of well spaced points the ROC curve's path can be traced with some confidence and the resulting value of $P(A)$ will be an unbiased estimate of the observer's sensitivity. As has been mentioned, $P(\bar{A})$ will give an unbiased sensitivity measure only if the observer has no bias either to signal or noise responses. With response bias $P(\bar{A})$ will underestimate sensitivity.

$P(c)_{2AFC}$ generally provides a good alternative measure of sensitivity to $P(A)$. It does not require the estimation of the path of the ROC curve and is thus efficient in respect of the number of experimental trials needed for its determination. There are occasions, however, when $P(c)_{2AFC}$ can underestimate sensitivity. In Table 2.11 is some data from a hypothetical observer in a 2AFC task. In Table

TABLE 2.11   *The effect of bias towards S responses in interval 1 of a 2AFC task on the value of $P(c)$ when sensitivity is held constant*

(a) No interval bias; $P(c) = \dfrac{40+40}{100} = 0.80$

		Observer's response to interval 1		
		S	N	Total
Stimulus in interval 1	s	40	10	50
	n	10	40	50
	Total	50	50	100

(b) Bias to S responses in interval 1; $P(c) = \dfrac{45+20}{100} = 0.65$

		Observer's response to interval 1		
		S	N	Total
Stimulus interval 1	s	45	5	50
	n	30	20	50
	Total	75	25	100

2.11(a) signal occurs fifty times in Interval 1 and also fifty times in Interval 2. It can be seen that $P(c) = 0.80$. Notice also that the column totals show that the observer makes the same number of signal responses to both intervals; that is he is not biased towards responding S to a particular interval. In Table 2.11(b) signal and noise again occur fifty times each in the two intervals. On this occasion the observer shows a strong bias towards seeing signals in interval 1, having moved half the responses in column N in Table 2.11(a) across into corresponding cells in column S. Although this is simply a change in bias (interval bias in this case) and sensitivity is unchanged, $P(c)$ drops to 0.65. Bias towards responding S in one interval can therefore artefactually depress $P(c)$ scores although there will be few examples of real experiments where this bias is as extreme as the illustration in Table 2.11(b) or where such a tendency in an observer cannot be rectified during practice trials.

Problems

1. With signal and noise events drawn from the same underlying distributions, an observer participates in three separate yes–no tasks. In the first he is instructed to be cautions in accepting events as signals; in the second he is told to use a medium degree of caution in detecting signals; and in the third he is told to adopt a lax criterion for signals.

The raw data for the experiment are given in the tables below. Determine the hit and false alarm rates for each criterion, plot the ROC curve, and from it find the observer's sensitivity using the area method.

Strict criterion			Medium criterion			Lax criterion		
	S	N		S	N		S	N
s	75	75	s	105	45	s	135	15
n	21	129	n	45	105	n	105	45

Responses made by an observer in three yes–no tasks with signal and noise held constant but with the criterion being varied.

47

2. Two observers participate in the same detection task and make their responses on a seven-point rating scale. From their raw data, given below, find:

(a) $P(S|s)$ and $P(S|n)$ for each rating category for each observer.

(b) From the ROC curves, by using $P(A)$, which of the two observers has the higher sensitivity.

(c) $P(\bar{A})$ for each criterion for observer 1. Compare the $P(\bar{A})$ scores with observer 1's $P(A)$ value. If $P(A)$ is a true estimate of the area under the ROC curve, what conclusions can you draw about errors which could occur in using $P(\bar{A})$ as a sensitivity measure?

| | | Rating categories Certain signal to certain noise | | | | | | | |
		1	2	3	4	5	6	7	Total
Observer 1	s	31	19	19	8	16	5	2	100
	n	2	5	9	7	27	19	31	100
Observer 2	s	23	21	6	10	9	22	9	100
	n	11	15	5	9	10	30	20	100

3. x is a variable which can take values from 1 to 10 in discrete steps of 1. In a detection task ten criteria are set up as follows:

For criterion a the observer responds S when $x \geqslant 1$.
For criterion b the observer responds S when $x \geqslant 2$.
For criterion c the observer responds S when $x \geqslant 3$.

.
.
.

For criterion j the observer responds S when $x = 10$.

With these 10 criteria the following hit and false alarm rates are observed:

Criterion	a	b	c	d	e	f	g	h	i	j
$P(S|s)$	1·00	1·00	0·94	0·87	0·74	0·60	0·40	0·26	0·13	0·06
$P(S|n)$	1·00	0·94	0·87	0·74	0·60	0·40	0·26	0·13	0·06	0·00

Determine $P(x|s)$ and $P(x|n)$ for each value of x and sketch the histograms for the signal and noise distributions.

4. Each of 12 observers works under both of two experimental conditions and produces the following hit and false alarm rates in yes–no tasks:

Observer	Condition A		Condition B	
	$P(S\|s)$	$P(S\|n)$	$P(S\|s)$	$P(S\|n)$
1	0·90	0·25	0·15	0·10
2	0·70	0·50	0·35	0·30
3	0·35	0·05	0·75	0·70
4	0·55	0·40	0·70	0·65
5	0·75	0·25	0·65	0·40
6	0·65	0·35	0·60	0·40
7	0·35	0·05	0·55	0·45
8	0·10	0·05	0·80	0·60
9	0·45	0·10	0·80	0·75
10	0·85	0·45	0·80	0·10
11	0·90	0·65	0·55	0·45
12	0·95	0·75	0·15	0·15

Use the area method illustrated in Figure 2.4 to decide, for each observer, whether sensitivity in condition A is better than (B), worse than (W), or indistinguishable (U) from sensitivity in condition B.

You may also like to use a sign test to decide whether sensitivity differs significantly between the two conditions.

5. For a variable x, the probability of x taking values between 1 and 6 in signal and noise trials is given below. From a matrix of joint probabilities of occurrence of pairs of signal and noise events, determine the proportion of correct responses expected in a 2AFC task. What other method might you have used to find $P(c)$?

x	1	2	3	4	5	6
$P(x\|s)$	0	1/9	2/9	3/9	2/9	1/9
$P(x\|n)$	1/9	2/9	3/9	2/9	1/9	0

49

Chapter 3

GAUSSIAN DISTRIBUTIONS OF
SIGNAL AND NOISE WITH
EQUAL VARIANCES

Although we have talked about distributions of signal and noise nothing has been said about the shapes they might assume. The sensitivity measures described in Chapter 1 were non-parametric and made no assumptions about the underlying distributions. There is, of course, no *a priori* reason that these distributions should be of any particular kind. Their shapes may vary capriciously from experiment to experiment. However if it did turn out that sensory events were distributed in the same way, this would allow very efficient sensitivity measures to be used, as will soon be seen. Most psychological variables seem to be at least approximately normally distributed and the first question that can be asked about the signal and noise distributions is, do they conform to a normal, or Gaussian, distribution?

In a detection task we do not directly observe the underlying signal and noise distributions. The data that we have consists of sets of hit and false alarm rates and it is from these that the underlying distributions must be inferred. We therefore wish to know whether there is anything which particularly distinguishes a set of hit and false alarm rates from Gaussian distributions from those of any other distribution.

THE ROC CURVE FOR THE YES–NO TASK
(Green & Swets, 1966, 58–62)

Let us consider the simplest example of a signal and a noise distribution both of which are distributed normally. They are shown in Figure 3.1. The x-axis shows the value of the evidence variable, x,

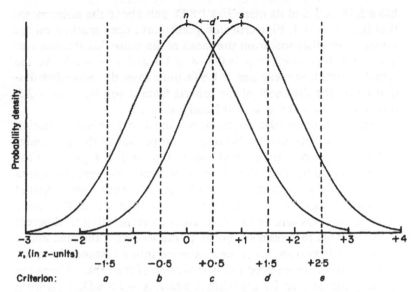

FIGURE 3.1 *Gaussian distributions of signal and noise. Noise mean = 0; signal mean = + 1. Noise variance = signal variance = 1*

and is calibrated in z, or standard deviation units of the distributions. The y-axis is marked 'Probability density' which may puzzle some readers. You will have noticed that all the histograms in Chapters 1 and 2 had their y-axes labelled 'Probability' and that this simply indicated the probability that x would take a particular value. 'Probability density' means much the same thing. It is just that 'probability' is conventionally used to refer to the y-axes of discontinuous distributions, i.e. those in which x takes a discrete set of values $(1, 2, 3, \ldots)$ etc. with no intermediate values between 1 and 2, 2 and 3, etc.) and the height of the distribution curve makes a series of discrete jumps (such as the bars on a histogram). On the other hand, it is conventional to use 'probability density' to refer to the y-axes of continuous distributions; that is where x takes a continuous range of values and the height of the distribution curve rises or falls in a continuously smooth curve.

In the figure, the noise distribution has a mean of 0 and a standard deviation of 1. Normal distributions with zero mean and unit S.D. are called *standard normal distributions*. The signal distribution also

51

has a S.D. = 1 and its mean lies 1 S.D. unit above the noise mean, that is, at $x = +1$. Five criterion points have been marked on the x-axis. Their distances from the mean of the noise distribution are: $a = -1.5$, $b = -0.5$, $c = +0.5$, $d = 1.5$ and $e = +2.5$. As the signal distribution mean lies $+1$ S.D. units from the noise distribution mean, the distances of the criteria from it will be: $a = -2.5$, $b = -1.5$, $c = -0.5$, $d = +0.5$ and $e = +1.5$.

Next, the hit and false alarm rates associated with each criterion can be found. This involves finding the proportion of the area under a standard normal curve which lies to the right of a point on the x-axis x S.D. units from the distribution mean. Normal curve area tables, with which you will already be familiar, have been prepared for doing this job. A set of these is given in Appendix 5. The tables come in two versions. In the first, areas corresponding to different distances from the mean are given. In the second, if the area above or below x is known, it can be converted into a distance from the mean. In this case we need the first version of the tables. Using the tables it can be seen for criterion a, which is -1.5 S.D. units from the noise distribution mean, 0.93 of the distribution lies above this point. Therefore $P(S \mid n) = 0.93$. Also a is -2.5 S.D. units from the signal distribution mean so, according to the normal curve tables, 0.99 of the signal distribution lies above a, meaning that $P(S \mid s) = 0.99$ for this criterion. Performing the same conversions for the remaining criteria, Table 3.1, which shows hit and false alarm rates for each criterion, is obtained.

TABLE 3.1 *The distance, in S.D. units, of the criteria in Figure 3.1 from the means of the signal and noise distributions and the hit and false alarm rates associated with each criterion*

| Criterion | Distance of criterion in S.D. units from: | | $P(S \mid n)$ | $P(S \mid s)$ |
	Noise mean	Signal mean		
a	−1.5	−2.5	0.93	0.99
b	−0.5	−1.5	0.69	0.93
c	+0.5	−0.5	0.31	0.69
d	+1.5	+0.5	0.07	0.31
e	+2.5	+1.5	0.01	0.07

With this set of hit and false alarm rates, the ROC curve for the signal and noise distributions of Figure 3.1 can be plotted. It is shown in Figure 3.2 and is a smooth curve running from (0, 0) to (1, 1). An ROC curve of this type is always obtained from Gaussian distributions of signal and noise with equal variances. Of course the curve will be bowed out to a greater or lesser extent depending on the amount of overlap of the two distributions, just as was seen in Chapter 2.

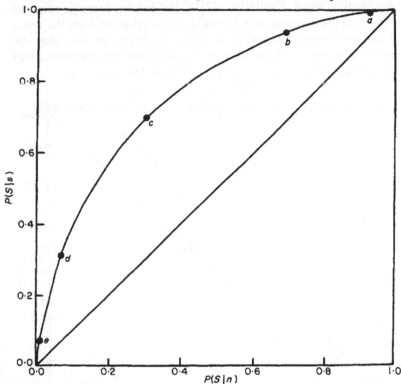

FIGURE 3.2   ROC *curve for Gaussian distributions of signal and noise of equal variance with the signal distribution mean one standard deviation above the noise distribution mean*

DOUBLE-PROBABILITY SCALES
(Green & Swets, 1966, 61)

So far, ROC curves have been plotted with the x- and y-axes having equal spaces between successive $P(S\,|\,s)$ or $P(S\,|\,n)$ values. It has just

53

been seen that any hit probability can be also represented as a distance of the criterion which gave rise to that hit probability from the mean of the signal distribution. Also any false alarm probability can be converted into a distance of the criterion from the noise distribution mean. Thus in place of $P(S|s)$ we could use the score $z(S|s)$ which is the criterion's distance in z-units (that is standard deviation units for a standard normal distribution) from the signal distribution mean. Similarly, $z(S|n)$, the z-score corresponding to $P(S|n)$, is the distance, in S.D. units, of the criterion from the noise distribution mean. Referring to Table 3.1 it can be seen that the scores in column 1 of the table are $z(S|n)$ values for the criteria and the scores in column 2 are the $z(S|s)$ values for the criteria.

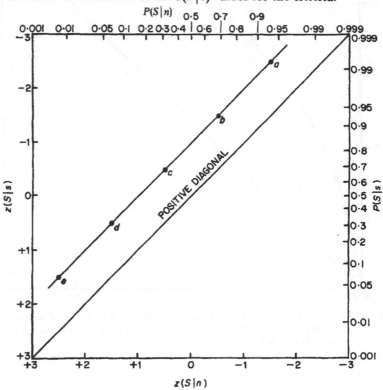

FIGURE 3.3 *Double-probability plot of the* ROC *curve for Gaussian distributions of signal and noise of equal variance and with the signal distribution mean one standard deviation above the noise distribution mean*

Another ROC curve can be constructed in which the x- and y-axes have equal spacings between successive z values. Such a scale is called a *double-probability scale*. On this $z(S \mid s)$ is plotted as a function of $z(S \mid n)$ as is shown in Figure 3.3. For the sake of comparison both P and z values are shown on the axes so that you can see that the ROC curve of Figure 3.3 is really the same as that of Figure 3.2 except that different scales have been used on the x- and y-axes. Notice also that as z is positive when $P < 0.50$ and negative when $P > 0.50$, the signs on the x- and y-axes are in the reverse directions to those conventionally used.

When the five pairs of $z(S \mid s)$ and $z(S \mid n)$ values have been plotted, it is found that they no longer give a curved ROC curve but one which is a straight line running parallel to the positive diagonal (i.e. the line passing through $(+3, +3)$ and $(-3, -3)$. An important principle can now be stated.

If the signal and noise distributions are Gaussian, they will give a curved ROC curve on a scale with equal spacings between successive P-values, but the ROC curve will be a straight line when plotted on a double-probability scale.

In a psychological experiment we usually have no *a priori* knowledge of the nature of the signal and noise distributions. What is observed is the hit and false alarm rates of an observer using a number of criteria. We can, however, work backwards from these and infer the nature of the signal and noise distributions by plotting the ROC curve on a double-probability scale, and if this turns out to be a straight line, it is likely that the underlying distributions were Gaussian.

Why should the double-probability plot be a straight line, and why should the line run parallel to the positive diagonal? Remember that $z(S \mid s)$ and $z(S \mid n)$ are distances from the means of two distributions with equal variances. Thus, in Figure 3.1, as we move 1 S.D. unit along the x-axis of the noise distribution we will also cover a distance of 1 S.D. along the x-axis of the signal distribution, that is to say, an increase of x for $z(S \mid n)$ will also result in an increase of x for $z(S \mid s)$. Consequently, $z(S \mid s)$ is linearly related to $z(S \mid n)$ and as both increase at the same rate the slope of the line will be equal to 1. Notice also that each $z(S \mid s)$ value in Table 3.1 is exactly 1 S.D. less than its corresponding $z(S \mid n)$ value. This relationship is, of

55

course, also apparent in the ROC curve of Figure 3.3. If any value of $z(S|s)$ is selected on the y-axis it will be found that the corresponding value of $z(S|n)$ on the x-axis is 1 more. This difference is due to the fact that the mean of the signal distribution, (from which point the $z(S|s)$ distances are measured) is 1 S.D. unit further up the x-axis than the mean of the noise distribution (from which the $z(S|n)$ distances are measured). It is therefore possible to find the distance between the distribution means from the double probability plot of an ROC curve for Gaussian distributions of signal and noise whose variances are equal. It is simply, for any criterion, $z(S|n) - z(S|s)$.

In signal detection tasks it is conventional to define the noise distribution mean as zero. An important measure of sensitivity d' (pronounced 'd-prime') can now be defined.

d' is the value of the signal distribution mean, measured in S.D. units of the noise distribution, when the noise distribution mean is equal to zero and both distributions are Gaussian and have S.D. $= 1$.

The measure d' is an alternative way of getting a sensitivity index from the method of measuring the area under the ROC curve described in Chapter 2. Both $P(A)$ and d' are indications of the extent to which the signal and noise distributions overlap and therefore both measure the same thing. However. d' involves a number of assumptions that $P(A)$ does not. First, both signal and noise must be distributed normally. This assumption appears to hold good for most experimental tasks. Second. both distributions must have the same variance. Often this appears to be the case but more exceptions to this rule are being found by experimenters. What one does when variances are not equal will be discussed in Chapter 4. and for the present it will be assumed that the second assumption can be held.

It was seen that whatever value of $z(S|s)$ was selected in Figure 3.3. the corresponding value of $z(S|n)$ was always 1 S.D. unit larger. Thus in this case five criterion points were not needed to construct the ROC curve. Any one of them would have sufficed. If the observer had used only criterion a, the single pair of values of $P(S|s)$ and $P(S|n)$ would have been enough to construct the entire ROC curve based on the equal variance assumption. For criterion a. $z(S|n) - z(S|s) = +1$. The ROC curve must therefore be a straight line passing through $z(S|s) = -2.5$, $z(S|n) = -1.5$ and with a slope of 1. d', the sensitivity measure, is $+1$.

If it is safe to make these assumptions about the nature of the signal and noise distributions a considerable saving can be made in experimental time. With one yes–no matrix, based on a single criterion, one hit and one false alarm rate is obtained. These are converted to z-scores by using normal curve tables and the difference $z(S|n) - z(S|s)$ is d', the sensitivity index.

THE FORMULA FOR d'

What has been said in the previous section can now be summarized in a more formal way. The sensitivity measure d' is the distance between the mean, $\overline{X}_s$, of a signal distribution and the mean, $\overline{X}_n$, of a noise distribution. So we can write:

$$d' = \overline{X}_s - \overline{X}_n. \qquad (3.1(a))$$

In the previous example $\overline{X}_n$ was equal to 0 and $\overline{X}_s$ to 1. Thus from (3.1(a)), $d' = 1 - 0 = 1$.

Looking back at the definition of d' it can be seen that it says d' is measured in standard deviation units of the noise distribution. To achieve this we need to ensure that $\overline{X}_s$ and $\overline{X}_n$ are themselves stated in noise S.D. units. To make this explicit in the formula for d', (3.1(a)) should be changed to read

$$d' = \frac{\overline{X}_s - \overline{X}_n}{\sigma_n}. \qquad (3.1(b))$$

where σ_n is the standard deviation of the noise distribution.

In the example of Figure 3.1 the distributions were drawn on an x-axis with the noise S.D. equal to 1 so that either version of (3.1) would have given the same answer.

You may notice the similarity between the formula for d' and the familiar formula used to convert a raw score, X, into a standard score (or a standard normal deviate). Standard scores are, of course, commonly used in psychological tests. If the score X came from a population of scores with mean μ and S.D. $= \sigma$ then z, the standard

57

score for X is given by

$$z = \frac{X - \mu}{\sigma}. \tag{3.2}$$

Formula (3.1b) is quite analogous to (3.2) and d' is just like a z-score, standard score or standard normal deviate.

In practice, as has already been seen, the working formula for finding d' from a pair of $P(S \mid s)$ and $P(S \mid n)$ values is

$$d' = z(S \mid n) - z(S \mid s) \tag{3.3}$$

where $z(S \mid s)$ is the z-score corresponding to $P(S \mid s)$, and $z(S \mid n)$ is the z-score corresponding to $P(S \mid n)$.

THE CRITERION

Just as it is possible to work backwards from an ROC curve to find the distance between the distribution means, so also can the curve be used to find the criterion points the observer used. Not all psychological experiments are interested in the measurement of sensitivity alone. Often measures of response bias are wanted as well.

Assume that an experiment has obtained from an observer the five pairs of $P(S \mid s)$ and $P(S \mid n)$ values which appear in Table 3.1. The experimenter first wishes to know what positions the criteria occupy on the x-axis upon which the signal and noise distributions are scaled. This is tantamount to asking what value of x corresponds to each criterion point. As Figure 3.1 shows, x is merely the distance of a point from the mean (of zero) of the noise distribution. It is already known that this distance, measured in noise standard deviation units, is the value of $z(S \mid n)$, the z-score corresponding to the $P(S \mid n)$ value for each criterion. So, by using normal curve tables to convert the $P(S \mid n)$ scores into $z(S \mid n)$ scores each criterion can be located in terms of its x value.

However, although finding the x value for each criterion allows its position to be determined relative to the means of the two distributions, x itself is not a direct measure of the extent to which a criterion represents a degree of bias towards either S or N responses. The score which will provide such information about bias is β, the criterion value of the likelihood ratio, which was introduced in Chapter 1.

58

To refresh your mind about the way in which β is calculated you may wish to look back at Figure 1.1. If an observer decides to respond S whenever $x \geqslant 66$ in. and to respond N whenever $x < 66$ in. the value of β will be $P(x|s)/P(x|n) = \frac{3}{4}$. It can be seen in Figure 1.1 that $P(x|s)$ and $P(x|n)$ are actually the heights of the signal and noise distributions at $x = 66$ in. So, if the height of the signal distribution is called y_s, and the height of the noise distribution is called y_n, an alternative way of writing formula (1.1), the expression for the likelihood ratio is

$$l(x) = y_s/y_n.\tag{3.4}$$

It can be checked that this way of writing the formula for the likelihood ratio will give $\beta = \frac{3}{4}$ for $x - 66$ in. in Figure 1.1. At $x = 66$ in., y_s is three units high and y_n is four units high. Substituting these values in (3.4) gives the correct β value.

Figure 1.1 is an example of two discrete distributions but the method for finding β is similar for continuous distributions. Figure 3.4 shows a pair of signal and noise distributions which are Gaussian. The distance between their means is equal to d', and a criterion, C, has been placed at a distance of $+x$ from the noise distribution mean. The value of β associated with this criterion will be given by dividing y_s, the height of the signal distribution at C by y_n, the height of the noise distribution at C. All that is needed is some means of finding these two heights. This can be done from the formula for the standard normal distribution curve. For the height of the noise distribution the formula is

$$y_n = \frac{e^{-\frac{1}{2}x^2}}{\sqrt{(2\pi)}},\tag{3.5a}$$

Where $x =$ the distance of the criterion from the mean of the noise distribution; in other words $z(S|n)$.

$\pi =$ the mathematical constant with the approximate value of $3\cdot142$.

$e =$ the mathematical constant with the approximate value of $2\cdot718$.

For the height of the signal distribution the formula is

$$y_s = \frac{e^{-\frac{1}{2}(x-d')^2}}{\sqrt{(2\pi)}}\tag{3.5b}$$

59

where $(x-d')$ is the distance of the criterion from the mean of the signal distribution, or, in other words, $z(S|s)$, and π and e are as for formula (3.5a).

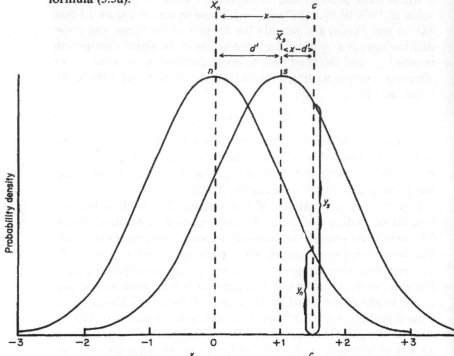

FIGURE 3.4 *Gaussian distributions of signal and noise and a single criterion, c. The height of the signal distribution is y_s at c, and the height of the noise distribution is y_n, $\bar{x}_s$, the signal distribution mean is at a distance d' above x_n, the noise distribution mean. c's distance from $\bar{x}_n$ is x, and from x_s, is $x-d'$*

As an illustration β will be calculated for criterion b of Table 3.1. We start by finding the height of the signal distribution at b.

The hit rate at b is seen to be 0·93. From the normal curve tables in Appendix 5 the $P(S|s)$ value of 0·93 gives $z(S|s) = -1·5$. As $z(S|s) = x-d'$ we can substitute in (3.5b) to get

$$y_s = \frac{2·718^{-\frac{1}{2}(-1·5)^2}}{\sqrt{(2 \times 3·142)}}$$
$$= 0·130.$$

60

Next the height of the noise distribution at b can be found. The false alarm rate at b is 0.69. From normal curve area tables $P(S \mid n) = 0.69$ gives $z(S \mid n) = -0.5$. As $z(S \mid n) = x$ we can substitute in (3.5a) to get

$$y_n = \frac{2.718^{-\frac{1}{2}(-0.5)^2}}{\sqrt{(2 \times 3.142)}}$$

$= 0.352$.

The final step is to divide y_s by y_n.

$$\beta = \frac{y_s}{y_n} = \frac{0.130}{0.352} = 0.37.$$

In practice it is not necessary to go through the labour of calculating y_s and y_n. The normal curve tables in Appendix 5 not only give P to z conversions but also the height of a standard normal curve for the range of P values. To find β, then, involves using the tables to find the heights (or ordinates as they are often called) of the distributions for the $P(S \mid s)$ and $P(S \mid n)$ values and then dividing the former height by the latter to obtain β. In Table 3.2 y_s, y_n and β have been found for each of the five criteria in Table 3.1. By looking at Table 3.1 and at Figure 3.1 which shows the criteria drawn against the signal and noise distributions it can be seen that β has the following properties.

First, at criterion c, which lies at the point where the signal and noise distributions intersect, β will be equal to 1. At this criterion, which lies midway between the means of the signal and noise distributions, the observer is no more biased to S than to N responses.

Second, below c the noise distribution lies above the signal distribution and β is less than 1. The further a criterion is removed from c the smaller β becomes. In Table 3.2 the β value for criterion a is less than that for b, and a lies further below c, the point of no bias, than b. The region below c also corresponds to criteria which show a bias towards S responses and the further below c the criterion is set, the greater the proportion of S responses made.

Third, above criterion c the signal distribution lies above the noise distribution and β will always be greater than 1. Also the region above c corresponds to a bias in favour of N responses.

Thus β gives a convenient measure of the response bias of any criterion point. If an observer's criterion gives $\beta = 1$, he is unbiased. If $\beta > 1$ the observer has adopted a strict or cautious criterion and is biased towards N responses. If $\beta < 1$ the observer has a lax or risky criterion and is biased towards saying S.

TABLE 3.2   *Calculation of β values for the criteria in Table* 3.1

Criterion		$x - d' =$ $z(S \mid s)$	Height of signal distribution, y_s.	$x =$ $z(S \mid n)$	Height of noise distribution, y_n.	$\beta = \dfrac{y_s}{y_n}$
Bias to signal.	a	$-2\cdot5$	$0\cdot018$	$-1\cdot5$	$0\cdot130$	$0\cdot14$
	b	$-1\cdot5$	$0\cdot130$	$-0\cdot5$	$0\cdot352$	$0\cdot37$
No bias	c	$-0\cdot5$	$0\cdot352$	$+0\cdot5$	$0\cdot352$	$1\cdot00$
Bias to noise.	d	$+0\cdot5$	$0\cdot352$	$+1\cdot5$	$0\cdot130$	$2\cdot71$
	e	$+1\cdot5$	$0\cdot130$	$+2\cdot5$	$0\cdot018$	$7\cdot22$

The meaure β can also be used in a more precise way as it is an expression of the observer's criterion in terms of the odds, for a particular value of x, of x being evidence for a signal. In Chapter 1 it was seen that for different probabilities of signal and noise and different stimulus-response outcomes, a β value could be calculated which would result in the maximizing of correct responses, or of the observer's net gain. An experiment wishing to investigate the effects of expectancy or pay-off on performance may involve the prediction of an ideal β value from formula (1.4) and comparison of this with β values obtained from real observers.

The range of values that β can take is sometimes inconvenient if response bias for a set of criteria is to be represented graphically. Criteria which represent biases towards S responses are restricted to the narrow range $0 < \beta < 1$ while criteria which represent biases towards N responses can take any value of $\beta > 1$. This can also lead to misinterpretation of the degrees of bias represented by βs. For instance $\beta = 2$ represents the same degree of bias to N as $\beta = 0\cdot5$ represents to S, while $\beta = 100$ represents the same degree of bias to N as $\beta = 0\cdot01$ represents to S. To equalize the intervals between degrees of response bias and to facilitate graphical representation of results, it is common practice to give bias scores in terms of $\log \beta$ rather than β itself. In Figure 3.5 the bias values for the five criteria of Table 3.2 are plotted against their criteria. It can be seen that the plot of $\log \beta$ gives a different, and a more meaningful picture of the degree of bias associated with each criterion than does the plot of β itself. When there is a bias to S responses $\log \beta$ will be negative. When there is a bias to N responses $\log \beta$ will be positive, and when bias is absent, $\log \beta$ will be zero.

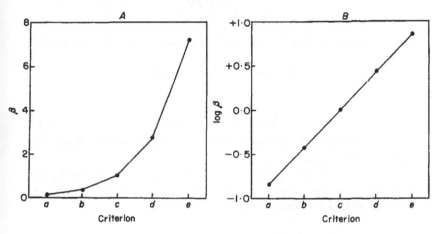

While $\log \beta$ can be found by using normal curve tables to find y_s and y_n, dividing y_s by y_n and taking the logarithm of the result, there is another means by which it can be calculated. This method involves the use of logarithms to the base e and readers who wish to revise their knowledge of these logarithms should consult Appendix 2 before going on with the rest of this section.

In this book the following conventions will be used for writing the logarithm of a number x; if the logarithm to the base 10 of x is being used, it will be written as $\log x$: if the logarithm to the base e of x is being used, it will be written as $\ln x$.

In Figure 3.4 a pair of Gaussian distributions of signal and noise were shown with the distance between their means being equal to d'. A criterion, C, was set at a distance x from the mean of the noise distribution. We have already seen that the value of β for criterion C is found by dividing (3.5(b)), the height of the signal distribution

63

at C, by (3.5(a)), the height of the noise distribution at C. Therefore the following expression can be written for β:

$$\beta = \frac{y_s}{y_n} = \frac{e^{-\frac{1}{2}(x-d')^2}/\sqrt{(2\pi)}}{e^{-\frac{1}{2}x^2}/\sqrt{(2\pi)}}.$$

Cancelling out $1\sqrt{(2\pi)}$ we are left with

$$\beta = \frac{e^{-\frac{1}{2}(x-d')^2}}{e^{-\frac{1}{2}x^2}}$$

Now, knowing that $\ln e^x = x$, the formula for $\ln \beta$ can be written

$$\ln \beta = \ln \frac{e^{-\frac{1}{2}(x-d')^2}}{e^{-\frac{1}{2}x^2}} = \ln e^{\frac{1}{2}d'(2x-d')}$$

so that:

$$\ln \beta = \tfrac{1}{2} d'(2x - d') \tag{3.6}$$

If it is desired $\ln \beta$ can be converted to $\log \beta$ by the formula: $\log \beta = 0.4343 \ln \beta$.

Formula (3.6) thus gives a convenient way of calculating an observer's bias from d' and $z(S \mid n)$. Tables of normal curve ordinates are unnecessary for this method.

FORCED-CHOICE TASKS
(Green & Swets, 1966, 64–9)

The forced-choice task was described in Chapter 2. It was shown there that $P(c)$, the proportion of correct responses made in the 2AFC task was the same as $P(A)$, the area under the equivalent yes–no task's ROC curve. It has just been seen that if Gaussian distributions of signal and noise are assumed it is possible to deduce, from points on the yes–no ROC curve, the distance between the means of signal and noise distributions. In this way the sensitivity measure d' was derived. If the same Gaussian assumptions are made about the under-lying signal and noise distributions used in a forced-choice task, it should also be possible to use data from a forced-choice experiment to estimate d'. This can be done both for 2AFC and mAFC tasks. For the 2AFC task the procedure for getting from $P(c)$ to d' is quite straightforward. For the mAFC task the calculations are somewhat more involved. Both types of task will be considered in turn.

*Finding d' in the 2*AFC *task*

The model for the forced-choice task which will be used here is one in which the observer has been presented with a stimulus item and is later shown a pair of test items. One of these is correct; it is the same as the stimulus and will be called signal. The other is incorrect; it differs from the stimulus and will be called noise. The observer is required to say which of the two test items is the signal.

Adopting the same assumptions as for the yes–no task the distribution of x for the noise item will be Gaussian with a mean of zero and S.D. $= 1$. The distribution of x for the signal item will also be Gaussian with a mean of d' and with S.D. $= 1$. The distributions are shown in Figure 3.6(a).

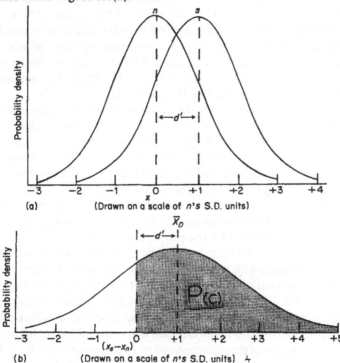

FIGURE 6.6   (a) *The distribution of the evidence variable x for a correct item. s, and an incorrect item. n, in a 2* AFC *task. Both distributions have S.Ds of 1.* (b) *The distribution of differences. $x_s - x_n$ with mean $= d'$ and S.D. $= \sqrt{2}$. The shaded area is P(c). the proportion of correct responses*

65

In the recognition task the observer has two pieces of evidence to consider. One is a value of x (call it x_n) which has been drawn at random from the noise distribution. The other is a value of x (call it x_s) which has been drawn at random from the signal distribution. He must decide which of these two xs is more likely to have come from the signal distribution. As was seen in Chapter 2 the x_s values are larger than the x_n values on the average, so that the observer's best strategy will be to call the larger of the two xs a signal.

Again, as was seen in Chapter 2, the proportion of correct responses, $P(c)$, will be equal to the probability of x_s being larger than x_n. In Chapter 2 this probability was found by pairing each x_s with each x_n value, finding the joint probability of their occurrence, and summing the $P(x_s . x_n)$s for all the cases where x_s exceeded x_n. In that example however x was a discrete variable with a finite number of values. In Figure 3.6(a), x is a continuous variable which can assume an infinite number of values. Thus it is impossible to pair each x_s and x_n to find $P(x_s \cdot x_n)$. In any case, this is not necessary. There is a better way of finding $P(c)$ than this.

First of all let us reformulate the decision the observer has to make. If he has to find the occasions when x_s is greater than x_n, this is the same as saying that he is looking for occasions when $x_s - x_n$ is greater than 0. If all x_s values were paired with all x_n values it would be possible to construct a distribution of the $x_s - x_n$ values resulting from these pairings. The area under this distribution of differences which lies above the point $x_s - x_n = 0$ would be equal to $P(c)$, the proportion of correct responses in the 2AFC task. The area lying below $x_s - x_n = 0$ would be the proportion of incorrect responses.

Now let us recall a common statistical principle about distributions of differences with which you will probably be familiar. If the distribution of a variable x_1 has a mean of $\overline{X}_1$ and a standard deviation of σ_1, and if the distribution of a variable x_2 has a mean of $\overline{X}_2$ and a standard deviation of σ_2 then the mean of the distribution of $x_1 - x_2$ (the distribution of differences) will be equal to the difference between the means of the distributions of x_1 and x_2 which is $\overline{X}_1 - \overline{X}_2$. Also the variance of the distribution of differences will be equal to the sum of the variances of the two distributions which is $\sigma_1^2 + \sigma_2^2$.

As $\overline{X}_s$, the mean of the signal distribution, is known to be d', and as $\overline{X}_n$, the mean of the noise distribution, is known to be 0, then $\overline{X}_D$

the mean of the distribution of $x_s - x_n$ will be $\overline{X}_s - \overline{X}_n = d' - 0 = +d'$. Also as the standard deviation of the signal distribution is 1 and the standard deviation of the noise distribution is also 1, then σ_D, the standard deviation of the distribution of $x_s - x_n$ will be $\sqrt{(1^2 + 1^2)} = \sqrt{2}$. This distribution is shown in Figure 3.6(b). The shaded area to the right of $x_s - x_n = 0$ is $P(c)$, the proportion of correct responses in the 2AFC task.

$P(c)$ can be determined from an observer's performance in a 2AFC task. It is then a matter of working back from this score to d'. This involves using our knowledge of the shaded area under the curve of Figure 3.6(b) to find how far the point $x_s - x_n = 0$ lies below the mean of the distribution, as this distance is equal to d'. This should be able to be done with the aid of normal curve area tables but it should first be noticed that as the distribution of differences has its S.D. $= \sqrt{2}$ it is not a standard normal distribution. When the distribution is rescaled in standard normal form, the distance from the mean to the point $x_s - x_n = 0$ is no longer d' but d'/σ_D which is equal to $d'/\sqrt{2}$. Now from $P(c)$ normal curve tables can be used to find the z-score, $z(c)$, above which the area $P(c)$ lies. Thus we have

$$z(c) = d'/\sqrt{2} \tag{3.7a}$$

and with a little re-arrangement,

$$d' = \sqrt{2} \cdot z(c). \tag{3.7b}$$

The relationship between forced-choice d' and yes–no d'

In a yes–no task the observer is given a single piece of evidence and must decide whether it was s or n. If both distributions are Gaussian and with standard deviations of 1 the distance between their means is d'.

In a 2AFC task the observer is given two pieces of evidence and must decide which was s and which was n. As Figure 3.7 shows, either of two things can happen. On some trials item 1 in the recognition test may be the signal and item 2, noise. For the sake of brevity such trials will be designated $\langle sn \rangle$. On other trials item 1 will be noise and item 2, signal. These trials will be called $\langle ns \rangle$.

It can be seen that while an observer in a yes–no task has to decide between the hypotheses s and n, the observer in the 2AFC task must decide between the two hypotheses $\langle sn \rangle$ and $\langle ns \rangle$. On any trial in

67

a 2AFC task the observer will have two pieces of evidence, one from the distribution for item 1 which will have a value of x_1, and the other from the distribution of item 2 which will have a value of x_2. Assume, as in the previous section, that the observer computes the difference $x_1 - x_2$ and if the difference is positive he decides in favour of hypothesis $\langle sn \rangle$ but if the difference is negative he decides in favour of $\langle ns \rangle$.

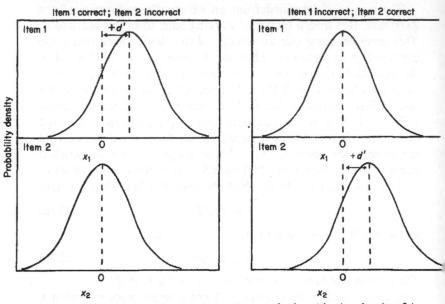

FIGURE 3.7 *Distribution of* x_1 *and* x_2 *in the* 2 AFC *task when either item 1 or item 2 is the correct item*

If $\langle sn \rangle$ is true it was seen in the preceding section that the distribution of $x_1 - x_2$ (which is of course the distribution of $(x_s - x_n)$) will have a mean of $+d'$ and a standard deviation of $\sqrt{2}$. On the other hand if $\langle ns \rangle$ is true the distribution of $x_1 - x_2$ (which is the distribution of $(x_n - x_s)$) will have a mean of $-d'$ and a standard deviation of $\sqrt{2}$.

These two distribution of differences are shown in Figure 3.8. They can be rescaled as standard normal distributions so that the mean of distribution $\langle sn \rangle$ becomes $+d'/\sqrt{2}$, and the mean of distribution $\langle ns \rangle$ becomes $-d'/\sqrt{2}$. In the yes–no task the sensitivity

68

measure was d', the distance between the means of s and n distributions. In an analogous manner the sensitivity measure for the forced-choice task which will be called d'_{FC} can be defined as the distance between the means of the distributions $\langle sn \rangle$ and $\langle ns \rangle$. This distance can be seen to be $2d'/\sqrt{2}$ or $d'\sqrt{2}$ where d' is the yes–no sensitivity measure. The relationship between the yes–no d' and d'_{FC} can therefore be written as

$$d'_{FC} = d'\sqrt{2} \qquad (3.8)$$

So for the same levels of signal and noise a forced-choice experiment will give a d' which is $\sqrt{2}$ larger than the d' obtained from a yes–no task. This statement must be immediately qualified by pointing out that this relationship between d'_{FC} and the yes–no d' will only hold if the two pieces of evidence, x_1 and x_2 are uncorrelated. In all further discussion it will always be assumed that the values of x occupying intervals in a forced-choice task have been chosen independently and are uncorrelated. That is not to say the assumption is true; it is merely convenient.

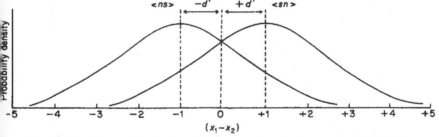

FIGURE 3.8   *The two distributions of differences for the 2 AFC task obtained from the distributions of item 1 and item 2 show in Figure 3.7. Distribution $\langle ns \rangle$ is obtained from $x_1 - x_2$ when item 2 is correct, and distribution $\langle sn \rangle$ is obtained from $x_1 - x_2$ when item 1 is correct*

However, Swets (1959) has shown empirically that d' estimated by a yes–no task corresponds closely to d' estimated by both 2AFC and 4AFC tasks, so we may accept the assumption of independence without many qualms.

The forced-choice ROC curve and forced-choice rating tasks
(Green & Swets, 1966, 68)

A 2AFC task can be converted into a rating scale task in much the

69

same way as a yes–no task can. Rather than presenting the observer with a pair of items and asking him to say which of them was the signal it is possible to make him indicate on a rating scale the degree of preference he has that the one item is a signal and the other noise. A four-point scale may used in the following way:

Category 1: Very certain that item 1 was signal,
Category 2: Moderately certain that item 1 was signal,
Category 3: Moderately certain that item 2 was signal,
Category 4: Very certain that item 2 was signal.

The raw data can be converted into two sets of cumulative probabilities as was illustrated for the yes–no task. However these probabilities mean different things in the two tasks. In the yes–no

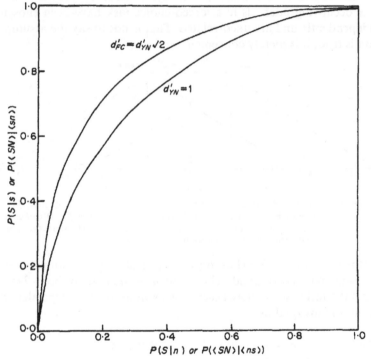

FIGURE 3.9 *Comparison of* ROC *curves for the yes–no and* 2 AFC *tasks. The yes–no d', d_{YN}, has been set equal to 1 so that $d_{FC} = 1\cdot41$*

70

task the observer was choosing between the two hypotheses s and n and the cumulative probabilities were values of $P(S \mid s)$ and $P(S \mid n)$. In the 2AFC task the two hypotheses are $\langle sn \rangle$ and $\langle ns \rangle$ so that the cumulative probabilities are values of $P(\langle SN \rangle \mid \langle sn \rangle)$ and $P(\langle SN \rangle \mid \langle ns \rangle)$. These two conditional probabilities are the areas to the right of a criterion which is being moved across the two distributions of $\langle sn \rangle$ and $\langle ns \rangle$ illustrated in Figure 3.8. The distributions are Gaussian and of equal variance, and when drawn as standard normal distributions have a distance of $d'\sqrt{2}$ between their means. The forced-choice ROC curve will thus be equivalent to a yes–no ROC curve for $d'\sqrt{2}$ as is shown in Figure 3.9.

The reader may wonder what advantages there are, if any, to be gained from estimating d'_{FC} from rating scale data rather than obtaining $P(c)$ and substituting in formula (3.7). In fact the conversion of $P(c)$ into d'_{FC} is even easier than this. Elliot (1964) has published tables of d' for forced-choice tasks not only for the 2AFC task but for a number of mAFC tasks where d' can be read off directly from $P(c)$. However, using $P(c)$ to obtain d' may result in an underestimation of sensitivity if the observer exhibits a bias towards calling one item in the recognition test 'signal' more often than the other. This so-called interval bias was discussed in the last section of Chapter 2, where it was found that $P(c)$ could, with a biased observer, underestimate $P(A)$, the area under the yes–no ROC curve. What was true for $P(A)$ is also true for d'. Green & Swets (1966, 408–11) describe a method for correcting $P(c)$ when interval bias is evident, but the other alternative is to use rating scale data to obtain the 2AFC ROC curve. Like the yes–no task the forced-choice curve can be plotted on a double-probability scale and d'_{FC} read off in the normal fashion.

The m-alternative forced-choice task
(Elliott, 1964; Green & Birdsall, 1964)

Not all forced-choice tasks are restricted to using two alternatives from which the subject must choose his response. One experiment by Miller, Heise and Lichten (1951) has used as many as 256 items from which the observer had to select the correct one. In Chapter 2 it was pointed out that it is often reported that $P(c)$ becomes smaller as m, the number of alternatives in the recognition task increases.

This section will give some idea as to why this effect occurs and, also, discuss the principles involved in obtaining d' from mAFC data.

Figure 3.10 illustrates the distribution of signal and noise for the mAFC task with one correct item (the signal) and $m-1$ incorrect items (all noise). The situation is much the same as for the 2AFC task. The signal distribution is Gaussian with S.D. $= 1$ and has a mean of d'. The noise distributions, $n_1, n_2, n_3, \ldots, n_{m-1}$ are also Gaussian with S.D.s of 1 and have means of 0. Again we assume that

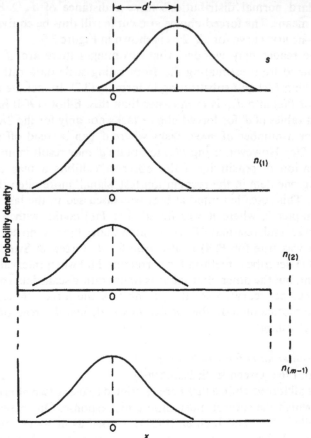

FIGURE 3.10 *Distributions of x for the mAFC task. There is one correct item, s, with a mean of d', and m−1 incorrect items. $n_1, n_2, \ldots, n_{m-1}$, with means of zero*

72

the observer selects the item which he will call 'signal' by choosing that one which has the largest value of x. In this case he will have a total of m values of x to consider: $x_s, x_{n(1)}, x_{n(2)}, x_{n(3)}, \ldots, x_{n(m-1)}$. Consider first the xs for all of the $m-1$ noise distributions. Each is a random drawing of an x from a distribution of mean 0 and S.D. $= 1$ so that the $m-1$ x_ns will cover a range of different values. Their average value will, of course be 0. One of these x_ns will be larger than the others and it will be denoted by the special symbol x_{max}. In the mAFC task, the observer will correctly select the signal item as often as its value, x_s, is larger than x_{max}, the largest of the x_ns.

Remembering the way in which $P(c)$ was estimated from the distribution of $x_s - x_n$ for the 2AFC task, it can be seen that for the mAFC task we need the distribution of differences between x_s and x_{max}. The first step, of course, is to find the distribution of x_{max} itself. This distribution is called Tippett's distribution and is approximately normal. It has a mean $\overline{X}_N$ which increases as the number of noise distributions in the mAFC task increases, and a standard deviation, σ_N, which decreases as the number of noise distributions increases.

It is easy to find $\overline{X}_N$ and σ_N for the 2AFC task as there is only one noise distribution. There the distribution of x_{max} is merely the distribution of x_n, that is, the noise distribution itself. Therefore in the 2AFC task, $\overline{X}_N = \overline{X}_n = 0$, and $\sigma_N = \sigma_n = 1$. For more than two alternatives the situation is a little more complicated and Tippett (1925) has prepared sets of tables to assist in the calculation of $\overline{X}_N$ and σ_N for distributions of x_{max} based on two or more noise distributions.

The mean of the distribution of differences between x_s and x_{max} will simply be $\overline{X}_s - \overline{X}_N$. $\overline{X}_s$ is known to be d', and $\overline{X}_N$ can be determined from Tippett's (1925) tables. So the mean of the distribution of $x_s - x_{max}$ is $d' - \overline{X}_N$. Also the standard deviation of the distribution of differences is $\sqrt{(\sigma_s^2 + \sigma_N^2)}$, and as $\sigma_s = 1$, the S.D. of the distribution of $x_s - x_{max} = \sqrt{(1 + \sigma_N^2)}$. The distribution of differences is illustrated in Figure 3.11. The shaded area to the right of $x_s - x_{max} = 0$ is $P(c)$, the proportion of correct responses in the mAFC task.

In Figure 3.11 it can be seen that the size of $P(c)$ depends on the size of $\overline{X}_D$, the mean of the distribution of differences. In turn $\overline{X}_D$

73

depends on the size of d', the signal distribution mean, and $\overline{X}_N$, the mean of the distribution of x_{max}. Previously it was said that as the number of noise distributions in the mAFC task increases, the mean of x_{max} also increases. This means that as the number of alternatives in the mAFC task goes up d' remains the same, and $\overline{X}_N$ goes up so that $\overline{X}_D$ will go down. Consequently $P(c)$ will become smaller. This is what has been observed in forced-choice tasks.

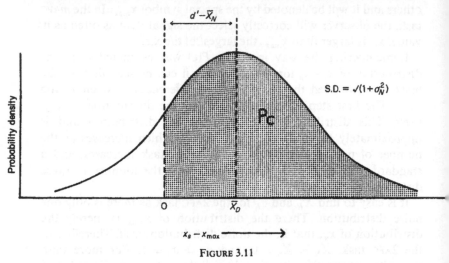

FIGURE 3.11

Although d' may not be changed by increasing the number of incorrect alternatives in an mAFC task, $P(c)$ will be decreased. Now this is not the complete story as not only does $\overline{X}_N$ increase with the number of noise distributions but σ_N also decreases. To understand the combined effects of these changes we need to write the formula for $\overline{X}_D$ in standard normal form which is

$$\overline{X}_D = \frac{d' - \overline{X}_N}{\sqrt{(1 + \sigma_N^2)}}. \tag{3.9}$$

As has been said, an increase in the number of noise distributions will give an increase in $\overline{X}_N$ so that the numerator of Formula 3.9 will become smaller. At the same time σ_N will become smaller so that the denominator of the formula will also become smaller. Whether $\overline{X}_D$, and hence $P(c)$, will decline, remain the same, or

74

increase as the number of noise distributions increases will therefore depend on the relative sizes of the changes in $\bar{X}_N$ and σ_N. In fact the change in $\bar{X}_N$ is not counterbalanced by the change in σ_N so that the net result in (3.9) is for $\bar{X}_D$ to decrease in size.

In the 2AFC task it is known that $\bar{X}_N = 0$ and that $\sigma_N = 1$. If these values are substituted in (3.9) it becomes

$$\bar{X}_{D(2\text{AFC})} = \frac{d' - 0}{\sqrt{(1+1)}} = \frac{d'}{\sqrt{2}}.$$

This last expression is the formula for $z(c)$ in the 2AFC task given in (3.7(a)). Thus (3.9) is just a more general version of (3.7(a)).

In practice it is not necessary to find $\bar{X}_N$ and σ_N before being able to convert $P(c)$ from an mAFC task into d'. Elliot's (1964) forced-choice tables give the conversion direct from $P(c)$ to d' for a number of mAFC tasks. However the discussion in this section should have made it clear that if an experiment has obtained $P(c)$ in a number of conditions where the number of incorrect alternatives in the forced-choice tasks has also been varied, it is not possible to compare sensitivity between these conditions simply by looking at differences in $P(c)$ values. As $P(c)$ varies as the number of alternatives vary it must first be converted into d' using Elliot's tables; d' is independent of the number of alternatives used in the recognition task so that sensitivity comparisons can now be made.

Problems

1. Find d' for the following hit and false alarm rates:

	$P(S\|s)$	$P(S\|n)$
(a)	0·27	0·13
(b)	0·58	0·21
(c)	0·85	0·62

2. Find d' for the following z-scores:

	$z(S\|s)$	$z(S\|n)$
(a)	−2·26	−0·95
(b)	+0·04	+1·76
(c)	−3·00	+0·39

75

3. Find d' when:
   (a) $P(N|s) = 0.80$ and $P(S|n) = 0.10$,
   (b) $P(N|n) = 0.57$ and $P(S|s) = 0.76$.

4. If:
   (a) $x = +1$ and $d' = 2$, what is $P(S|s)$?
   (b) $z(S|s) = 0.5$ and $d' = 0.75$, what is $P(S|n)$?
   (c) $x = +0.25$ and $z(S|s) = -0.55$, what is d'?
   (d) $d' = 1$ and $x = +1.5$, what is $P(N|s)$?
   (e) $P(N|n) = 0.73$ and $x - d' = -0.25$, what is d'?

5. Which of the following pairs of hit and false alarm rates represent equivalent levels of sensitivity?
   (a) $P(S|s) = 0.46$ and $P(S|n) = 0.26$,
   (b) $P(S|s) = 0.84$ and $P(S|n) = 0.31$,
   (c) $P(S|s) = 0.93$ and $P(S|n) = 0.50$,
   (d) $P(S|s) = 0.82$ and $P(S|n) = 0.47$.

6. Find β for the three pairs of hit and false alarm rates in problem 1.

7. With the aid of formula (3.6) find $\ln \beta$ from the three pairs of z-scores in problem 2.

8. In a yes–no task an observer has a false alarm rate of 0.2 and $\beta = 1$. Find:
   (a) d',   (b) $P(S|s)$.

9. Observers participate in a task in which they hear either bursts of noise or noise plus a weak tone. They are told to identify trials on which tones occurred. The observers are allocated to two groups. Group 1 receives placebos which, it is said, will enhance ability to detect tones. Group 2 receives the same placebos and is told that they will decrease sensitivity to tones.

The experimenter has two hypotheses:
(a) That group 1 will pay more attention to the display than group 2 and thus show superior discrimination of the tones.

(b) That group 1 will be more willing to report tones than group 2 but that discrimination will not differ between the two groups.

If the following results are typical of observers in the two groups, which hypothesis appears to be supported by the data?

		Group 1 Number of responses		Group 2 Number of responses	
		Tone present	Tone absent	Tone present	Tone absent
Number of stimuli	Tone present	172	28	104	96
	Tone absent	108	92	36	164

10. From the following rating-scale data for a yes–no task find the values of $z(S|s)$ and $z(S|n)$ for each rating category, plot the ROC curve on a double-probability scale and determine the value of d'.

		High certainty signal		to	High certainty noise	
		1	2	3	4	5
Stimuli	s	264	132	60	102	42
	n	30	54	42	162	312

Observer's response

11. In a 2AFC task an observer makes 165 correct responses out of a possible 220. Find:

(a) The yes-no d' corresponding to this performance,

(b) d'_{FC}

77

12. In a 2AFC rating-task the following raw data are obtained:

		High certainty item 1 correct	to		High certainty Item 2 correct
		1	2	3	4
Stimulus condition	Item 1 correct	46	36	10	8
	Item 2 correct	7	24	19	50

Plot the forced-choice ROC curve on a double-probability scale and determine d'_{FC} and the yes–no d' corresponding to it.

Chapter 4

GAUSSIAN DISTRIBUTIONS OF SIGNAL AND NOISE WITH UNEQUAL VARIANCES

Unfortunately for the experimenter, not all detection tasks involve signal and noise distributions with equal variances. In these cases we cannot infer the entire ROC curve from a single pair of hit and false alarm probabilities but need a set of points for the curve. This means that while a single yes–no task is sufficient to estimate d' in the equal variance case, a series of yes–no tasks, or a rating scale task, is necessary to determine sensitivity when variances are not equal.

Given that the distributions of signal and noise are Gaussian, there are four parameters associated with the two distributions which can vary. These are the mean and variance of the signal distribution and the mean and the variance of the noise distribution. All four of these cannot be measured simultaneously. In Chapter 2, to define d', the noise distribution mean was fixed at the arbitrary value of 0. Similarly, to measure the variance of the signal distribution, the noise distribution's variance must also be held constant. It is customary to assume that the variance of the noise distribution is 1. It must be noted that fixing the noise mean at zero and the noise variance at unity is quite arbitrary; any other pair of values might have been used. The advantage in using this particular mean and this particular variance is that the noise distribution is thus defined as a standard normal one which makes subsequent calculations simpler. Having defined the noise distribution in this way, the things that will be able to be measured from experimental data are the signal distribution mean, the signal distribution variance and the criterion, either in terms of x or β.

79

ROC CURVES FOR UNEQUAL VARIANCE CASES

Curves plotted on scales of $P(S \mid s)$ and $P(S \mid n)$
(Green & Swets, 1966, 62–4)

In Figures 4.1 and 4.2 two examples of unequal variance are shown. In Figure 4.1 the variance of the signal distribution is less than that of the noise distribution. In Figure 4.2 the variance of the signal distribution is greater than that of the noise distribution. Had we plotted $P(S \mid s)$ as a function of $P(S \mid n)$ we would have obtained the ROC curves shown alongside each pair of distributions. These curves are not of the simple type obtained for the equal variance case illustrated in Figure 3.2. In Figure 4.1. the ROC curve generated by moving a criterion from left to right across the two distributions

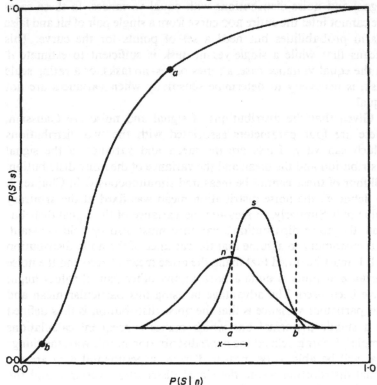

FIGURE 4.1 *The general shape of the* ROC *curve when* σ_s *is less than* σ_n.

80

starts at $P(S|s) = P(S|n) = 1$ and bows out until it reaches the point a. The curve then bows inwards until b is reached. The shape of this curve reflects the rate at which $P(S|s)$ changes over the range of values of $P(S|n)$, and the points of maximum outward and inward deflection correspond to the two positions on the x-axis of the distributions where the curves of the signal and noise distributions intersect. This means that there will be two points where $\beta = 1$. The hit and false alarm rates for $\beta = 1$ will be those for points a and b on the ROC curve.

Figure 4.2 behaves similarly to Figure 4.1 except that, beginning at $P(S|s) = P(S|n) = 1$, it bows inwards to c and then out to d. The points c and d also give a β value of 1.

Had an experiment yielded ROC curves of the types shown in

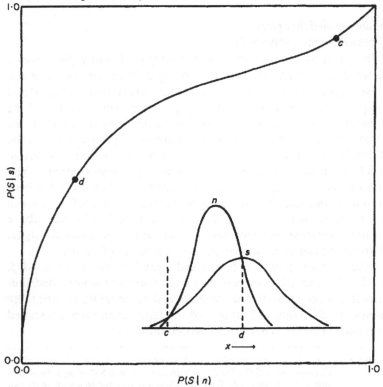

FIGURE 4.2   *The general shape of the* ROC *curve when* σ_s *is greater than* σ_n.

81

Figures 4.1 and 4.2, we might have inferred that the variances of the signal and noise distributions were different. If the ROC curve had been like the one in Figure 4.1, this would have suggested that $\sigma_s < \sigma_n$. Had the ROC curve been like that one in Figure 4.2, this would have suggested that $\sigma_s > \sigma_n$.

In fact it is quite unlikely that real experimental data would ever produce such well-defined curves. Points b and c lie very close to the ends of their ROC curves. In the case of b this is a region where $P(S|s)$ and $P(S|n)$ have very small values. The region on which c lies represents very large values of $P(S|s)$ and $P(S|n)$. It is very difficult to accurately estimate hit and false alarm rates for criteria in the extreme tails of the signal and noise distributions, nor do observers often oblige by adopting criteria of such strictness or laxity.

Double-probability plots
(Green & Swets, 1966, 96–7)

In the unequal variance case a double-probability plot gives a better idea of the nature of the underlying distributions of signal and noise than does the ROC curve scaled in equal intervals of $P(S|s)$ and $P(S|n)$ units. Figure 4.3 is an example where the variance of the signal distribution is greater than that of the noise distribution. To avoid confusion, both distributions have been plotted on separate scales. Look first at the x-axis of the noise distribution. It is scaled in S.D. units of the noise distribution as in previous examples. You can see that the S.D. of the noise distribution has therefore been fixed at 1, and also that the mean of the distribution is 0. The mean of the signal distribution, $\bar{X}_s$, can be seen to be 1 noise S.D. above the noise distribution mean, and the standard deviation of the signal distribution can be seen to be equal to 1·5 noise S.D. units.

Five criterion points, a, b, c, d and e are located at $-1·5$, $-0·5$, $+0·5$, $+1·5$ and $+2·5$ respectively on the x-axis of the noise distribution. It is simply a matter of using normal curve tables to turn these values of $z(S|n)$ into $P(S|n)$ to find the false alarm rates associated with each criterion. They are shown in Table 4.1.

FIGURE 4.3 (page 83) *A signal distribution whose mean lies $+1$ noise S.D. units above the noise mean. S.D. of noise distribution = 1, and S.D. of signal distribution = 1·5 × noise S.D. Each distribution is depicted on a scale of its own S.D. units*

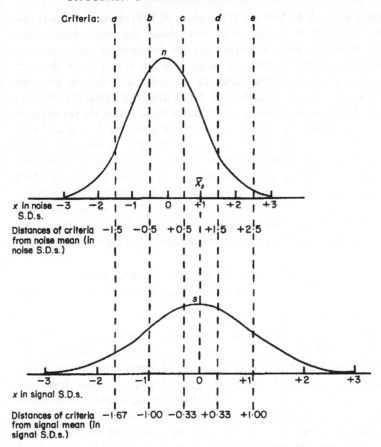

TABLE 4.1   *Calculation of hit and false alarm rates for the unequal variance case illustrated in Figure 4.3*

Criterion	$z(S\|n)$ Distance from noise mean in noise S.D.s	$z(S\|s)$ Distance from Signal mean in signal S.D.s	$P(S\|n)$	$P(S\|s)$
a	−1·50	−1·67	0·93	0·95
b	−0·50	−1·00	0·69	0·84
c	+0·50	−0·33	0·31	0·63
d	+1·50	+0·33	0·07	0·37
e	+2·50	+1·00	0·01	0·16

Next we wish to find the five values of $P(S|s)$. As the signal distribution mean is 1 noise S.D. unit above the noise distribution mean it will be easy to determine the distance of each criterion point from the mean of the signal distribution. It will be the distance of the criterion from the mean of the noise distribution plus the distance of the noise distribution mean from the signal distribution mean. As the signal mean is $+1$ from the noise mean, the noise mean will be -1 from the signal mean, so that points a, b, c, d and e lie at -2.5, -1.5, -0.5, $+0.5$ and $+1.5$ respectively from the signal mean. Now these distances are in noise S.D. units and the standard deviation of the signal distribution is not equal to the standard deviation of the noise distribution. Before normal-curve area tables can be used to convert the distances into $P(S|s)$ values, they will have to be transformed to standard scores by dividing each one by the standard deviation of the signal distribution. Thus we can write:

$$z(S|s) = \frac{\text{Distance of criterion from signal mean}}{\text{Size of a signal standard deviation}} \qquad (4.1)$$

both quantities in noise S.D. units.

To see how this works out in practice, criterion a is -2.5 noise S.D.s from the signal distribution mean. The size of a signal distribution standard deviation is equal to 1.5 noise distribution S.D.s. Therefore in standard normal form, $z(S|s)$, a's distance from the signal mean is $-2.5/1.5 = -1.67$. Performing similar calculations for b, c, d and e the values in column 2 of Table 4.1 are obtained.

Table 4.1 now contains the necessary information for finding $P(S|s)$ as this just requires using normal curve tables to convert $z(S|s)$ values into probabilities. This conversion is shown in the last column of the table. An ROC curve on scales of $P(S|s)$ and $P(S|n)$ can now be plotted (Figure 4.4) and although this curve resembles Figure 4.2 it is difficult to detect the rapid acceleration of the curve in the upper right-hand corner. However, if the curve is compared with Figure 3.2, the ROC curve for the equal variance case, it can be seen that while equal variance of signal and noise produce an ROC curve which is symmetrical, unequal variances give a characteristically asymmetrical curve.

What will Figure 4.4 look like on a double-probability plot?

Table 4.1 gives the necessary values of $z(S|s)$ and $z(S|n)$ for the criteria and $z(S|s)$ is plotted as a function of $z(S|n)$ in Figure 4.5.

Again, as in the equal variance case, the ROC curve is a straight line, but, whereas equal variances gave a line with a slope of 1 the slope of this curve is less than 1.

Some other important principles about ROC curves can now be be stated.

The ROC curve derived from distributions of signal and noise which are Gaussian will always be a straight line when plotted on a double-probability scale.

If the variances of the two distributions are equal this line will have a slope of 1.

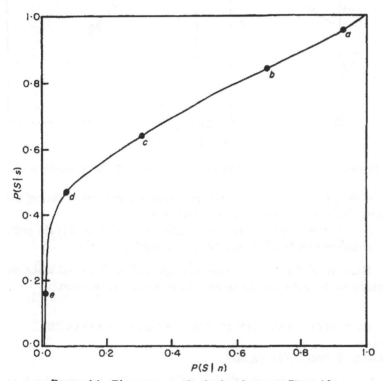

FIGURE 4.4 *The* ROC *curve for the distributions in Figure 4.3.*

85

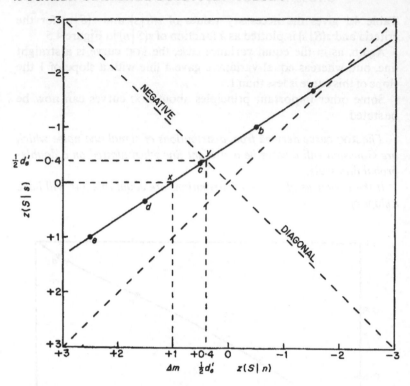

FIGURE 4.5 *The* ROC *curve of Figure 4.4 replotted on a double-probability scale.*

If the variance of the signal distribution is greater than that of the noise distribution the slope will be less than 1.

If the variance of the signal distribution is less than that of the noise distribution then the slope will be greater than 1.

The reasons for these changes in slope will be discussed when we return to the standard deviation of the signal distribution.

SENSITIVITY MEASURES IN THE UNEQUAL VARIANCE CASE
(Green & Swets, 1966, 96, 407)

From the ROC curve of Figure 4.5 is it possible to work back to the underlying distributions of signal and noise which gave rise to it? We found this to be possible with Figure 3.3, the double-probability plot of the ROC curve for the equal variance case. There, if we selected any value of $z(S|s)$ on the y-axis and read off the corresponding value of $z(S|n)$ on the n-axis, it was found that $z(S|n)$ was always 1 more than $z(S|s)$ and that this difference was d', the distance between the means of the two distributions. If we try to do this with Figure 4.5, we run into trouble. As the slope of the ROC curve is less than 1, the difference between $z(S|n)$ and $z(S|s)$ changes as we select different points along the ROC curve. Down near point e the difference between $z(S|n)$ and $z(S|s)$ is large and positive but becomes smaller as we move up the curve towards a. Just beyond a, where the curve crosses the positive diagonal, the difference is zero. After this it becomes negative. This illustrates the danger in trying to estimate d' from a single pair of hit and false alarm rates if there is a likelihood that the variance of the signal and noise distributions are not equal. Had we taken two observers, one of whom alway used criterion a and the other of whom alway used criterion e we would have obtained $d' = 0.17$ for the first observer and $d' = 1.50$ for the second. On the face of it, this constitutes quite a big difference in sensitivity, but in truth, both points lie on the same ROC curve and must therefore represent equivalent degrees of sensitivity but different degrees of bias. In Figure 4.5 it is just not possible to construct the whole ROC curve from a single pair of hit and false alarm probabilities unless we have some *a priori* information about the variance of the signal distribution. Of course, this is just the kind of information that we do not possess before conducting an experiment. The signal distribution has to be inferred from the shape of the ROC curve so this means obtaining several pairs of hits and false alarms for different criteria so that the curve's path can be traced.

When variances are unequal, we will therefore have to adopt some convention as to the point on the ROC curve at which the sensitivity measure is read. Commonly two conventions are used. The first has the virtue of being directly related to the distance between the distributions' means. The second has been shown empirically to give a reasonably stable estimate of the observer's sensitivity.

87

Δm as a measure of sensitivity

Formula (3.3) showed that the working formula for obtaining d' in the equal variance case was $d' = z(S|n) - z(S|s)$. We already know from Figure 4.3 that the distance between the means of the two distributions in this example is $+1$ noise S.D. If we examine the ROC curve in Figure 4.5 it can be seen that there is just one value of $z(S|s)$, which when substituted in the formula, will give $d' = +1$. This is when $z(S|s) = 0$ and at that point marked X on the ROC curve, $z(S|n)$ can be seen to be equal to $+1$. Only this pair of values will give the correct distance between the distribution means. The reason for this can be seen from the drawing of the distributions in Figure 4.3. The x-axis of the signal distribution is scaled in $z(S|s)$ values, and $z(S|s) = 0$ is the point at which the signal distribution mean lies. It can be seen that this is also the point $\bar{X}_s$ on the noise distribution's x-axis which is scaled in steps of $z(S|n)$. So, $\bar{X}_s$ lies $+1$ $z(S|n)$-unit from the mean of the noise distribution.

A new sensitivity measure, Δm, (pronounced 'Delta-em') can now be defined.

Δm is the distance between the means of the signal and noise distributions measured in standard deviation units of the noise distribution. It is equal to $z(S|n)$ at the point on the ROC curve where $z(S|s) = 0$.

If $\bar{X}_s$ is the signal distribution mean and $\bar{X}_n$ the noise distribution mean we can then write

$$\Delta m = \bar{X}_s - \bar{X}_n \qquad (4.2(a))$$

and, as the definition of Δm specifies that it is being measured in noise S.D.s, formula (4.2(a)) is more correctly written as

$$\Delta m = \frac{\bar{X}_s - \bar{X}_n}{\sigma_n} \qquad (4.2(b))$$

Formulae (4.2(a)) and (4.2(b)) are quite like formulae (3.1(a)) and (3.1(b)), the expressions for d'. Thus Δm and d' are equivalent measures but we keep d' to refer to the special case where the variances of signal and noise distributions are equal.

The working formula for Δm is also similar to formula 3.3, the working formula for d'. From the definition of Δm it follows that

$$\Delta m = z(S \mid n) - z(S \mid s)$$

at the point where

$$z(S \mid s) = 0. \qquad (4.3)$$

d'_e as a measure of sensitivity

Δm is not the only measure currently used as an index of sensitivity in the unequal variance case. Egan (see Egan & Clarke, 1966) has proposed another measure which he calls d'_e. (It is also sometimes called d_s). The convention here is to read the value of $z(S \mid s)$ or $z(S \mid n)$ at the point where the ROC curve, plotted on a double-probability scale, intersects the *negative diagonal*. As you can see in Figure 4.5, the negative diagonal is the straight line passing through $(0, 0)$ and the points $(+1, -1)$, $(-1, +1)$, etc. In Figure 4.5 the ROC curve meets the negative diagonal at the point y, and it will be noticed that at this point both $z(S \mid s)$ and $z(S \mid n)$ are equal to 0.4 (ignoring the differences in sign). d'_e *is defined as twice the value of* $z(S \mid s)$ *or* $z(S \mid n)$, *ignoring signs, at the point where the ROC curve intersects the negative diagonal*. In this case, therefore, $d'_e = 0.8$.

The fact that $z(S \mid s)$ and $z(S \mid n)$ are equal when the ROC meets the negative diagonal is one reason in favour of using d'_e as a measure of sensitivity. This is because it gives equal weight to units of both the signal and noise distributions. In the case of Δm sensitivity is scaled in units of the noise distribution. Thus this measure is appropriate if we expect noise variance to remain constant over a series of experimental treatments but for signal variance to change. If we suspect that both signal and noise variance may change over a series of treatments d'_e would be a more appropriate sensitivity measure. Also, if we expect signal variance to remain constant and noise variance to change, the most appropriate sensitivity measure would be the value of $z(S \mid s)$ at the point on the ROC curve where $z(S \mid n) = 0$. This sensitivity measure is scaled in units of the signal distribution and is employed in Thurstonian category scaling (Lee, 1969).

Another argument which makes d'_e a desirable measure of sensitivity is that the point from which it is read from the ROC curve normally falls within the range of points produced by the observer's criteria. Often the criteria will not give points which lie as far down the ROC curve as the point from which Δm is read. Consequently Δms point may have to be obtained by extrapolation. On the whole it is safer to use a measure which does not have to be extrapolated and d'_e usually meets this requirement.

Thirdly, Egan & Clarke (1966) report that the slopes of ROC curves tend to be unstable and to vary somewhat from session to session for the same observer. These changes in slope appear to alter the value of Δm more than the value of d'_e, which is the more stable measure.

All other things being equal, it does not really matter at what point the sensitivity measure is read from the ROC curve. If we decide to use Δm it is easy to find what d'_e would have been by using the conversion formula provided by Green & Swets (1966):

$$d'_e = 2\Delta m \left[\frac{s}{1+s} \right] \tag{4.4}$$

where s is the slope of the ROC curve.

Conversely, Δm can be calculated from d'_e by the formula

$$\Delta m = \frac{d'_e(1+s)}{2s}. \tag{4.5}$$

Although the reader may find Δm to be a more meaningful sensitivity index because it is related directly to the distance between the means of signal and noise distributions, he should not place too much store on this apparent reasonableness of the measure. Firstly, the point at which a sensitivity measure is taken from an ROC curve is quite arbitrary. As long as the convention is made explicit and adhered to in the experiment, comparisons of sensitivity measured under different conditions will probably be valid. Secondly, although we have spoken about ROC curves with slopes not equal to 1 as being the result of unequal variances of signal and noise, this is not inevitably the case. The slope of the curve can indicate other things as well; for example, distributions which are not truly Gaussian (see Lockhart & Murdock, 1970, for a discussion of this). Thirdly, if, as it seems, d'_e is a more stable measure than Δm, then it is preferable purely on these grounds.

90

MEASURING THE SIGNAL DISTRIBUTION VARIANCE

It has been seen that the slope, s, of the ROC curve can be used to tell us whether the variance of the signal distribution is less than, equal to, or greater than the variance of the noise distribution. It seems reasonable that the more the slope of the curve departs from 1 the greater the difference between the variances, but what is the exact relationship between s and signal variance?

In Chapter 3 we discussed why the ROC curve for the equal variance case was a straight line and why its slope was equal to 1. Briefly, the argument went like this: $z(S|s)$ and $z(S|n)$ are the distances of criteria from the two distribution means. Thus as we moved 1 S.D. unit along the x-axis on which the noise distribution was drawn, a distance of 1 S.D. was also covered along the x-axis of the signal distribution. Therefore an increase of x in $z(S|n)$ was accompanied by an increase of x in $z(S|s)$. Consequently a plot of $z(S|s)$ against $z(S|n)$ would be linear, and, as both increase at the same rate, the slope of the line would be 1.

The same argument applies for the unequal variance case. However, as Figure 4.3 shows, moving 1 noise S.D. along the x-axis of the noise distribution will result in a move of only $0 \cdot 67$ of a signal S.D. along the x-axis of the signal distribution. To give a specific example; moving from criterion d to criterion e involves going from $z(S|n) = +1 \cdot 5$ to $z(S|n) = +2 \cdot 5$, an increase of 1 $z(S|n)$-unit. Making the same move from d to e along the signal distribution's x-axis involves going from $z(S|s) = +0 \cdot 33$ to $z(S|s) = +1 \cdot 00$, an increase of $0 \cdot 67$ $z(S|s)$-units. To cover a distance of 1 S.D. along the x-axis of the signal distribution we would have to move $1 \cdot 5$ S.D.s along the x-axis of the noise distribution. The ROC curve in which $z(S|s)$ is plotted against $z(S|n)$ is still a straight line but it will have a slope of $0 \cdot 67/1$. This slope is, of course, equal to the ratio σ_n/σ_s. As the value of σ_n has been defined as 1 then s, the slope, is equal to $1/\sigma_s$. Thus the relationship between the slope of the ROC curve and the variance of the signal distribution can be stated as follows:

The reciprocal of the slope, s, of the ROC curve is equal to the standard deviation of the signal distribution.

This line of reasoning can be checked from the ROC curve itself.

91

In Figure 4.5 it can be seen that an increase of 1 unit along the x-axis will result in an increase of 0·67 of a unit along the y-axis so that the slope will be 0·67 which is what was predicted.

THE MEASUREMENT OF RESPONSE BIAS

It has already been seen that when the variances of the signal and noise distributions are not equal the curves of the two distributions will intersect twice resulting in two points where β is equal to 1.

If Figure 4.1 ($\sigma_n > \sigma_s$) is examined again it will be seen that from $P(S|s) = P(S|n) = 1$ to the point a, the noise distribution lies above the signal distribution so that in this region β will be less than 1. From a to b the signal distribution lies above the noise distribution so that β will be greater than 1. Beyond b the noise distribution lies above the signal distribution again so that β will be less than 1. An observer wishing to maximize his correct responses should therefore take heed of these changes in β. Ideally, if x, the evidence variable, has a value less than a he should respond N. If x lies between a and b he should respond S. But, if x is greater than b he should respond N again.

For Figure 4.2 ($\sigma_s > \sigma_n$) the situation is analogous to that in Figure 4.1. Below c, β exceeds 1; from c to d, β is less than 1; and above d, β is greater than 1 again.

Unlike the equal variance case, unequal variances produce β values which are not monotonic with the value of the evidence variable, x. That is to say, in the equal variance situation there was a simple relationship between x and β. If x increased then so did β. If x decreased, β values also went down. Unfortunately, this convenient state of affairs no longer applies when variances differ.

From a psychological point of view we would like to know what effects these changes in β will have on an observer. If, for example, signal variance is less than noise variance, does the observer stop responding S and start responding N when x exceeds b in Figure 4.1? Similarly, when $\sigma_s > \sigma_n$ does the observer respond S to values of x less than c in Figure 4.2? Unfortunately, there is currently no evidence which will give a clear answer to these questions. We have already seen that the estimation of points at the extreme ends of an ROC curve is an unreliable business, in addition to which most observers

92

do not adopt criteria of such strictness or laxity for them to fall into the critical regions where the β values reverse.

However, these questions do pose an important problem. On what basis do real observers in psychological experiments set up their response criteria? So far we have seen that the criterion which best achieves an observer's aims, whether they be the maximizing of correct responses, the minimizing of certain types of errors, or the maximizing of net gain, can be stated in terms of a likelihood ratio. That is, the selection of a criterion involves the choice of some appropriate value of β. Although this may be the most rational way of deciding whether to respond S or N to a particular piece of evidence, there is no inevitability that observers will actually use likelihood ratios to select criteria. It is not inconceivable that the criterion for response is based directly on the value of x, the evidence variable. Should this be the case, the observer will select a value of x as his criterion, and evidence which produces a sensory effect less than this value will be judged to be noise while evidence whose sensory effect exceeds this value will be judged to be signal. In the equal variance case there need not be any conflict between a decision rule based on x and one based on β as the one varies as a monotonic function of the other. However, in theory, discrepancies could occur when variances are unequal.

It is of little consolation to the potential user of detection theory that such a basic issue should be undecided. No mathematical model yet devised, however, has relieved a researcher of the obligation to interpret his own data, and it still rests with the experimenter to decide in his particular case what is the most appropriate criterion measure. From a practical point of view it would be both wise and helpful for experimenters not only to report the β values corresponding to their observers' criteria, but also the value of x for each criterion as well. As x is merely the value of $z(S \mid n)$ for each criterion, and as the noise distribution is always defined with a mean of 0 and an S.D. of 1, there is no difference between the calculation of this measure in the unequal variance case and its determination in the equal variance case. Some changes will need to be made in the calculation of β.

The basic principles involved in finding β in the unequal variance case are the same as those described in Chapter 3 for distributions of equal variance. For a given criterion point the $P(S|s)$ and $P(S|n)$ values are used to find y_s and y_n, the heights of the signal and noise distributions. y_s is then divided by y_n to obtain β. As the noise distribution is the same when variances are equal or unequal, there is no difference in the calculation of y_n in the two cases. The height of the noise distribution can still be found by using Formula (3.5a) or by using normal curve tables to find the ordinate corresponding to the $P(S|n)$ value for the criterion.

In the unequal variance case the formula for the height of the signal distribution needs to be modified to take account of the signal variance. The modified formula is

$$y_s = \frac{1}{\sigma_s\sqrt{(2\pi)}}\exp\left[\frac{(x-\Delta m)^2}{2\sigma_s^2}\right] \qquad (4.6)^1$$

This formula is still like (3.5(b)) which was used to calculate the height of a signal distribution with a standard deviation of 1. The following differences can be noticed:

(a) The term σ_s^2 now appears in (4.6). It could have been included in the same places in (3.5(b)) but, as the signal S.D. was equal to 1 in that formula, it was omitted.

(b) In place of the term $(x-d')$ in (3.5(b)) the term $(x-\Delta m)$ appears in (4.6). $(x-\Delta m)$ is the distance of the criterion from the mean of the signal distribution measured in S.D. units of the noise distribution. Now, looking back at (4.1), it can be seen that the expression for finding $z(S|s)$ could have been written symbolically as

$$z(S|s) = \frac{x-\Delta m}{\sigma_s}. \qquad (4.7)$$

The right-hand side of this expression appears in (4.6), so that the expression for the height of the signal distribution could be rewritten in terms of $z(S|s)$ in the following way:

$$y_s = \frac{1}{\sigma_s}\cdot\frac{1}{\sqrt{(2\pi)}}e^{-\frac{1}{2}z(S|s)^2} \qquad (4.8(a))$$

And, as $1/\sigma_s = s$, the slope of the ROC curve then:

$$y_s = s\left[\frac{1}{\sqrt{(2\pi)}}e^{-\frac{1}{2}z(S|s)^2}\right] \qquad (4.8(b))$$

---

[1] The term 'exp' in the equation denotes exponentiation, or the raising of e to some power. Thus another way of writing e^x would be $\exp[x]$

TABLE 4.2 *Determination of β values for the unequal variance case. The data is taken from Table 4*

Criterion	(1) $P(S\mid n)$	(2) $P(S\mid s)$	(3) y_n	(4) Ordinate for $z(S\mid s)$	$y_s = z$ ×	(6) ordinate (s)	$\beta = y_s$
a	0·93	0·95	0·13	0·10			0·54
b	0·69	0·84	0·35	0·24			0·46
c	0·31	0·62	0·35	0·38			0·71
d	0·07	0·47	0·13	0·38			1·52
e	0·01	0·16	0·03	0·24			5·33

It can be seen that the part of (4.8(b)) contained within the large square brackets is the expression for the height of a standard normal distribution at a point $z(S|s)$ standard deviations from its mean. Thus the ordinate of this distribution can be found from the normal curve tables in Appendix 5. All that is required is to find the ordinate for the $P(S|s)$ value of the criterion. To convert this ordinate into y_s one simply multiplies it by the slope of the ROC curve.

The data from Table 4.1 will now be used to illustrate the calculation of β for the unequal variance case. The following steps are involved:

(a) The values of $P(S|n)$ and $P(S|s)$ are listed in columns 1 and 2 of Table 4.2.

(b) Normal curve tables are used to find the ordinate of a standard normal curve corresponding to each $P(S|n)$ value. These are the values of y_n for each criterion and are shown in column 3 of the table.

(c) The normal curve tables are used to find the ordinate of a standard normal distribution curve corresponding to each $P(S|s)$ value. These are shown in column 4 of the table.

(d) Each of the ordinates found in (c) above is multiplied by s, the slope of the ROC curve, which in this case has been found to be 0.67. The resulting values, listed in column 5 of the table, are the values of y_s.

(e) As $\beta = y_s/y_n$, each entry in column 4 is now divided by its corresponding entry in column 2 to give the β values for each criterion. These are shown in column 6.

It can be seen that the β values do not increase as a simple function of x. The x values are given in column 7 for the sake of comparison.

Problems

1. Given a noise distribution of standard normal form and a signal distribution with a mean, $\overline{X}_s$, and a standard deviation of size σ_s noise S.D.s, use the following information to find the position of a criterion, C.

(a) Distance of C from $\overline{X}_s$ in noise S.D.s $= -1.0$. $\sigma_s = 1.4$. Find the distance of C from $\overline{X}_s$ in signal S.D.s.

(b) Distance of C from noise mean in noise S.D.s $= -0.8$.

96

Distance of X_s from noise mean in noise S.D.s = $+2\cdot2$. $\sigma_s = 0\cdot6$.
Find the distance of C from X_s in signal S.D.s.

(c) Distance of noise mean from X_s in signal S.D.s = $-1\cdot1$. Distance of C from X_s in signal S.D.s = $+0\cdot4$. $\sigma_s = 1\cdot5$. Find the distance of C from the noise mean in noise S.D.s.

2. If a point on an ROC curve is for $P(S|s) = 0\cdot65$ and $P(S|n) = 0\cdot14$, find Δm and d'_e for the following values of σ_s.
   (a) $\sigma_s = 0\cdot30$.   (b) $\sigma_s = 2\cdot35$.   (c) $\sigma_s = 1\cdot43$.

3. (a) If $\Delta m = 1\cdot5$ and $\sigma_s = 2$, what is $P(S|n)$ when $P(S|s) = 0\cdot84$?
   (b) If $d'_e = 1$ and $\sigma_s = 0\cdot67$, what is $P(S|s)$ when $P(S|n) = 0\cdot07$?

4. If a point on an ROC curve is for $P(S|s) = 0\cdot42$ and $P(S|n) = 0\cdot21$, find σ_s when Δm or d'_e have the following values:
   (a) $\Delta m = 0\cdot5$.   (c) $d'_e = 0\cdot2$.
   (b) $\Delta m = 0\cdot7$.   (d) $d'_e = 0\cdot8$.

5. (a) If $\Delta m = 1\cdot4$ and $s = 0\cdot4$, what is d'_e?
   (b) If $d'_e = 0\cdot12$ and $s = 2$, what is Δm?
   (c) If $\Delta m = 2$ and $d'_e = 0\cdot8$, what is s?

6. An experimenter obtains the following hit and false alarm rates from two observers:
   Observer 1: $P(S|s) = 0\cdot27$, $P(S|n) = 0\cdot04$.
   Observer 2: $P(S|s) = 0\cdot69$, $P(S|n) = 0\cdot42$.
   He knows that in both cases $\sigma_s/\sigma_n = 0\cdot8$. Do the observers show comparable levels of sensitivity?

7. Find β in the following cases where:
   (a) $P(S|s) = 0\cdot74$, $P(S|n) = 0\cdot17$ and $\sigma_s = 1\cdot25$.
   (b) $P(S|s) = 0\cdot38$, $P(S|n) = 0\cdot14$ and $s = 1\cdot4$.
   (c) $P(S|s) = 0\cdot50$, $P(S|n) = 0\cdot31$, $\Delta m = 1\cdot5$ and $d'_e = 1\cdot8$.
   (d) $x = +0\cdot45$, $\Delta m = 0\cdot20$ and $\sigma_s = 0\cdot75$.
   (e) $z(S|s) = -0\cdot55$, $z(S|n) = -0\cdot25$ and $s = 1\cdot6$.

97

8. In a rating-scale experiment the following raw data are obtained:

Observer's response
High certainty signal to high certainty noise

Category:		1	2	3	4	5
Stimulus	s	12	30	31	15	12
event:	n	5	9	10	10	66

Plot the ROC curve on a double-probability scale and find:

   (a) Δm (b) d'_e

   (c) σ_s.   (d) β for each criterion.

   (e) If one of the points where $\beta = 1$ lies approximately near criterion 1, can you determine, by sketching the distributions, the values of $z(S\,|\,s)$ and $z(S\,|\,n)$ for the other point where β is 1?

   (f) What will be the values of $P(S\,|\,s)$ and $P(S\,|\,n)$ at the point where the ROC curve intersects the curve for $d' = 2$?

Chapter 5

CONDUCTING A RATING SCALE EXPERIMENT

From the preceding chapters it should now be clear that the safest way of gaining information about an observer's sensitivity is by obtaining from him a number of values of $P(S|s)$ and $P(S|n)$ which will allow the path of the ROC curve to be determined. In Chapter 2 it was found that although a single pair of hit and false alarm rates could be used to make an approximate estimate of the area under the curve, the measure $P(\bar{A})$ was not independent of response bias and would underestimate sensitivity to a degree dependent on the observer's tendency to prefer S to N responses. In Chapter 4 it was seen that a single pair of hit and false alarm rates could give different values of d' for points on the same ROC curve if the slope of the curve was not equal to 1. In a 2AFC task, interval bias can affect the obtained value of $P(c)$ and hence, d'_{FC}.

Two methods have been described for obtaining a set of hit and false alarm rates which represent different degrees of response bias but equivalent degrees of sensitivity. The first required conducting a series of yes–no tasks with signal and noise strengths held constant but with the observer varying his bias from task to task. The second involved the use of a rating scale on which the observer simultaneously held several criteria of differing degrees of strictness. As the rating scale method is the most efficient as far as the number of trials needed to obtain the ROC curve points is concerned, it is likely to be the procedure most favoured by potential users of detection theory. In the following sections of this chapter, a rating scale task is worked through and a number of problems which an experimenter may encounter are discussed. The example used is a yes–no rating scale task but most of the steps described could be applied equally to a 2AFC rating task.

99

EXPERIMENTAL DESIGN

The nature of the experiment

The experiment to be described is one conducted by the author and concerned with short-term memory for sequences of digits. Bushke (1963) has devised a technique for assessing short-term memory called the 'missing scan'. In this task the subject hears a sequence of nine different digits from the series 0 to 9 and is required to name the missing tenth digit.

The missing scan technique was converted to a yes–no detection task by presenting ten digits to the observer. The first nine were the stimulus items and were nine different digits, in random order, from the series 0 to 9. The tenth digit was the test item. On half the trials it was the digit which had been omitted from the stimulus series, and on half the trials it was a repetition of one of the nine stimuli. The observer's task was to decide whether the test digit was a repetition of a stimulus digit or not.

Signal and noise trials

In this experiment a signal trial was defined as one in which the tenth digit was a repetition, and a noise trial as one in which the test digit was different from the nine stimuli.

Altogether there were 288 signal and 288 noise trials.

At this point the question may be raised: How many trials of signal and noise are necessary to get a good estimate of the ROC curve? Green & Swets (1966) suggest that somewhere in the region of 250 of each is desirable in a rating scale task. Normally, unless the experimenter is interested in the effects of *a priori* probabilities of signal and noise events on his observers' performances $P(s)$ and $P(n)$ are set at 0·5. There is nothing wrong with making signal trials twice as frequent as noise trials, or selecting any proportions of s and n that the experimenter desires. However, Ogilvie & Creelman (1968) point out that for a given number of total trials in an experiment, equal numbers of s and n trials give the most reliable estimate of the points for the ROC curve.

The prospect of having to run some 500 trials per experimental condition may be daunting to many researchers. A person who wishes to measure sensitivity and bias for each observer, under each of four conditions, has committed himself to 2000 trials per subject. If it takes about 15 seconds per trial and it is decided that each session will last for 1 hour, such an experiment would involve some eight 1 hour sessions per observer.

For many experiments these demands are impossible. One notable example is the case of an experimenter who, working in a mental hospital, wishes to conduct detection experiments with abnormal subjects. The ability of such subjects to tolerate long hours of testing is severely limited. No more than one session per week may be possible, and it is likely that many of the subjects will have undergone treatment during this time, which may vitiate the results. Another example is an experiment involving the administration of a drug, where testing of subjects must be done rapidly before the effects of the drug wear off.

It would be unfortunate if experimenters in areas such as these were debarred from using detection techniques. It will be suggested later in this chapter that if the experimenter is willing to compromise a little by using non-parametric measures of performance, such as $P(A)$, rather than parametric measures, such as Δm and d'_e, he can reduce his experiment to more realistic proportions. The author and his collaborators have found that by such means, tolerable measures of sensitivity are obtainable in some cases with as few as fifty signal and fifty noise trials. Realistically, a distinction must be made between experimenters who want precise parametric data to fit psychophysical functions and those who merely wish to know whether sensitivity and bias differ from one situation to another.

Choice of a rating scale

The next decision facing the experimenter concerns the number of categories to be used in the rating scale and the manner in which these should be described to the observer. The first thing to remember is that we will end up with one less useful point for the ROC curve than the number of rating categories we started out with. An experimenter who begins by deciding that he will use three rating categories will be left with only two points for his double-probability ROC curve. This is not only insufficient to determine the path of the curve visually, but also debars the experimenter from using any curve-fitting procedure when he comes to analyse his data. So, at a minimum, four rating categories should be used.

There are no hard and fast rules for choosing a maximum number of categories but the following points are worth remembering. (For an extended discussion of the ins and outs of rating scale methods, the reader should consult Guilford (1936).

First, although the more rating categories used, the more points available for the ROC curve, there is a risk that observers will not use a large number of categories very consistently. Probably ten categories can be used by an observer after a reasonable amount of practice, but there may be diminishing returns beyond this point (Conklin, 1923; Symonds, 1924).

Second, the greater the number of categories, the greater the chances that some will not be used by the observer. This may prove problematic if the observer fails to use categories at the extreme ends of the scale, as values of $P(S|n) = 0$ and $P(S|s) = 1$ may be obtained. When converted to z-scores such values will give $z(S|n) = -\infty$ and $z(S|s) = +\infty$ which are not easy to fit on the plot of the ROC curve.

Having said this, however, there is at least one experiment (Watson, Rilling & Bourbon, 1964) where observers indicated their confidence by moving a pointer along a scale, thus effectively having an infinite number of rating categories in which to put their responses. This method yielded a large number of points for the ROC curve, enabling its path to be traced with a high degree of accuracy. This method is most attractive to an experimenter who has an experimental set-up capable of logging rating scale data automatically. Not only can observers be inconsistent in using a large number of categories; experimenters have been known to misread scales when transcribing readings on to response sheets as well.

In general, if the experimenter wishes to use a large number of categories, he must also use a large number of signal and noise trials to ensure that $P(S|s)$ and $P(S|n)$ values associated with each category are accurately estimated. For most purposes four to ten categories will suffice. Also it is often permissible procedure for an experimenter, after seeing his data and deciding that too many categories were used, to combine adjacent ones so as to give more reliable data from which to calculate hit and false alarm rates.

102

Finally, it is important that the observer understands what the rating scale means. Any detection task requires some practice trials but even after the most careful precautions, some observers forget what they have been told, or being overwhelmed by the demands of the task, fall into confusion. Several methods of presenting the rating scale have been used. The first simply involves denoting each category by a number. The observer may be told that category 1 is for those events that he considers to be certainly signals, category 2 for events which are likely to be signals, category 3 for events which are likely to be noise, and category 4 for events which are certainly noise. This method has the virtue of simplicity from the experimenter's point of view, but observers have been known to forget whether '1' meant 'certain signal' or 'certain noise'.

The second method uses the following type of scheme:

'+ +' means 'certain the event was a signal'.
'+' means 'probably signal, but uncertain'.
'−' means 'probably noise, but uncertain'.
'− −' means 'certain the event was noise'.

This sort of scheme does appear to be more meaningful to observers and they are less likely to forget which ends of the scale refer to signal and noise than when using the numerical categories.

A third method is still less confusing for the subject. He makes his response in words, e.g.

'Certain—signal'
'Uncertain—signal'
'Uncertain—noise'
'Certain—noise'

A fourth procedure used by Swets, Tanner & Birdsall (1961) requires the observer to rate the stimulus according to the probability that it is a signal. Thus category 1 on the rating scale may indicate the probability of 0.00 to 0.04 that the stimulus was a signal, category 2, a probability of 0.05 to 0.19, and categories 3, 4, 5, and 6, probabilities of 0.20 to 0.39, 0.40 to 0.59, 0.60 to 0.79 and 0.80 to 1.00, respectively.

The experimental set-up will determine the form of the rating scale to some extent; for instance, whether the observer is to report

103

verbally, check a category on a response sheet, or press a button on a key-board. In the example used here, observers used a ten-point numerical scale, giving their reports verbally to the experimenter after each trial. Subsequently these ten categories were reduced to five by combining the data in category 1 with that in category 2, category 3 with category 4 and so on.

Structure of experimental trials

The way in which trials are conducted will vary according to the demands of the experiment but the most general paradigm for the sequence of events in a single trial is as follows:

Warning signal		Stimulus event		Test item		Observer's response		Feedback of information
	→		→		→		→	
(1)		(2)		(3)		(4)		(5)

In the example being used here stage 3, the test item, is distinct from the presentation of stage 2, the presentation of the stimulus. In other experiments where the subject knows what he is looking for, these two stages may occur concurrently, as for example when the observer is attempting to identify the presence of a tone embedded in a burst of white noise.

Often it is desirable to follow the observer's response by some feedback about how well he is performing. This may be a necessary feature of the experimental plan if the experimenter is interested in looking at changes in the response criterion as a function of different costs and rewards for various stimulus-response combinations. On the other hand the experimenter may give feedback of results purely to offset boredom and to keep the observer responding at a stable level. In such a case feedback may take the form of telling the subject whether the stimulus for that trial was signal or noise, or telling him whether his response was correct or incorrect, or merely giving a running total of correct responses made to that point in the experiment.

In other cases, such as in the example used here, feedback may be given during practice trials, as this usually allows the observer to settle down quickly to a stable level of performance, but dispensed with during the test trials themselves.

ANALYSIS OF DATA

Collation of raw data

After completing the experimental trials the first step in analysis is to separate signal from noise trials and to total these separately for the observer according to the rating response each trial received. The data for the example are presented in this form in Table 5.1.

TABLE 5.1 *The number of times an observer rated 288 signal and 288 noise trials in each of the five response categories*

Observer's response
High certainty signal to high certainty noise

Category	1	2	3	4	5	Total
Stimulus s	159	41	19	37	32	288
event n	2	3	21	80	182	288

Procedure for dealing with empty cells

After sorting out the raw data the experimenter may have to decide whether to combine some of the categories before going further with the analysis. Instead of the data in Table 5.1 he may

TABLE 5.2 *Variations of Table 5.1 but where the observer has failed to make either signal or noise responses in one of the rating categories*

Case 1. No noise response in category 1.

Observer's response
High certainty signal to high certainty noise

Category	1	2	3	4	5	Total
Stimulus s	159	41	19	37	32	288
event n	0	5	21	80	182	288

Case 2. No signal responses in category 5.

Observer's response
High certainty signal to high certainty noise

Category	1	2	3	4	5	Total
Stimulus s	159	41	19	69	0	288
event n	2	3	21	80	182	288

105

have obtained either of the two possibilities in Table 5.2. In the first case no occurrences of noise received the rating response '1', and in the second case no occurrence of signal received rating response '5'. The first case will result in $P(S \mid n)$ for category 1 being 0, and the second case will cause $P(S \mid s)$ for category 4 to be 1. Both these will give z-scores of infinity which will be of no use in determining the path of the ROC curve on a double-probability scale.

Two courses of action are open to the experimenter. He may either remove the offending categories by combining them with adjacent ones as has been done in Table 5.3, ot he may retain the categories and add to, or subtract from them a small arbitrary constant when he comes to calculating $P(S \mid s)$ and $P(S \mid n)$. This second procedure will be illustrated in the next section.

TABLE 5.3   *The data from Table 5.2 with categories combined to eliminate cells in which no responses occur*

Case 1. Categories 1 and 2 combined.

Observer's response
High certainty signal to high certainty noise

Category		(1 + 2)	3	4	5	Total
Stimulus	s	200	19	37	32	288
event	n	5	21	80	182	288

Case 2. Categories 4 and 5 combined.

Observer's response
High certainty signal to high certainty noise

Category		1	2	3	(4 + 5)	Total
Stimulus	s	159	41	19	69	288
event	n	2	3	21	262	288

Calculation of hit and false alarm probabilities

The next step is to convert the raw data into a set of hit and false alarm rates. The method for doing this was described in Chapter 2. This involved starting with the strictest 'signal' category (category 1) and collapsing the table into a yes–no matrix from which $P(S \mid s)$

and $P(S|n)$ were determined by dividing the number of hits and false alarms by their respective row totals (in this case 288). This calculation appears in Table 5.4.

TABLE 5.4 *Calculation of hit and false alarm probabilities for the raw data of Table 5.1*

Observer's response
High certainty signal to high certainty noise

Category	1	2	3	4	5
Sum of responses to signal in category i or stricter	$= 159$	$159 + 41$ $= 200$	$159 + 41 + 19$ $= 219$	$159 + 41$ $+ 19 + 37$ $= 256$	$159 + 41 + 19$ $+ 37 + 32$ $= 288$
$P(S|s) = \text{Sum}/288$	0·55	0·69	0·76	0·89	1·00
Sum of responses to noise in category i or stricter	$= 2$	$2 + 3$ $= 5$	$2 + 3 + 21$ $= 26$	$2 + 3$ $+ 21 + 80$ $= 106$	$2 + 3 + 21$ $+ 80 + 182$ $= 288$
$P(S|n) = \text{Sum}/288$	0·01	0·02	0·09	0·37	1·00

For the sake of comparison the hit and false alarm rates for the collapsed-category data in Table 5.3 are given in Table 5.5. Comparison of these hits and false alarms with Table 5.4 shows that they are the same except that in case 1, $P(S|s)$ and $P(S|n)$ are missing for category 1, while in case 2, there are no hit and false alarms for category 4. A moment's reflection about the method by which the raw data are cumulated to calculate $P(S|s)$ and $P(S|n)$ should make it clear why the collapsed data should give the same hits and false alarms as the uncollapsed data. Thus dealing with categories in the raw data table by combining them with adjacent ones does not result in any distortion of the values of $P(S|s)$ and $P(S|n)$ on which the ROC curve is to be based. The only penalty attached to the method is that an ROC curve point will be lost.

The alternative suggested in the preceding section was to retain all the categories regardless of whether they had been used by the observer or not, and to add a small constant to P-values which turned out to be 0, and to subtract a small constant from those

107

TABLE 5.5 *Hit and false alarm rates for the collapsed-category data of Table 5.3*

Case 1. Categories 1 and 2 combined.

Observer's response
High certainty signal to high certainty noise

Category	$(1+2)$	3	4	5
$P(S\mid s)$	0·69	0·76	0·89	1·00
$P(S\mid n)$	0·02	0·09	0·37	1·00

Case 2. Categories 4 and 5 combined.

Observer's response
High certainty signal to high certainty noise

Category	1	2	3	$(4+5)$
$P(S\mid s)$	0·55	0·69	0·76	1·00
$P(S\mid n)$	0·01	0·02	0·09	1·00

which turned out to be 1. A constant of 0·001 might be thought appropriate. This means that $P(S\mid n)$ for category 1 of case 1 in Table 5.2 would be set at 0·001 instead of its correct value of 0·01 while $P(S\mid s)$ for category 4 of case 2 in Table 5.2 would be set at 0·999 instead of its correct value of 0·89. This technique has much less to recommend it than the collapsing method. Although all points for the ROC curve are retained, those that have been saved by adding or subtracting arbitrary constants may be so much in error as to distort the curve when it is plotted. More will be said about the effects of $P(S\mid s)$ and $P(S\mid n)$ values which reach either 1 or 0 when we come to the sections on non-parametric measures of sensitivity and bias.

Conversion of $P(S\mid s)$ and $P(S\mid n)$ to $z(S\mid s)$ and $z(S\mid n)$

In order to obtain the points for the double-probability plot of the ROC curve, the values of $P(S\mid s)$ and $P(S\mid n)$ in Table 5.4 are converted to $z(S\mid s)$ and $z(S\mid n)$ by the use of the normal curve area tables in Appendix 5. These z-scores are shown in Table 5.6.

Determination of the path of the ROC curve

In Figure 5.1 $z(S\mid s)$ is plotted against $z(S\mid n)$. The next task is to

TABLE 5.6 *The $z(S|s)$ and $z(S|n)$ values for the $P(S|s)$ and $P(S|n)$ values of Table 5.4*

	Observer's response High certainty signal to high certainty noise				
Category	1	2	3	4	5
$z(S|s)$	−0·13	−0·50	−0·71	−1·23	+∞
$z(S|n)$	+2·32	+2·05	+1·34	+0·33	+∞

decide which straight line passes through the four points. In previous examples, points of the ROC curve were chosen to fall exactly along a straight line. In the present example, as is often the case with data from real experiments, there is no straight line which will pass exactly through all the points. Error in the estimation of hit and false alarm rates may result in points being randomly displaced from their ROC curve. What we need to do is to choose a line so that all the points lie as close to it as possible.

In many textbooks on statistics, a procedure called the *method of least squares* is given for fitting a straight line to a number of points. This method normally involves the selection of a line which minimises the *vertical* distances of the points from the line. The method is illustrated in Figure 5.2. One of the assumptions underlying the method is that the random displacements of the points are solely due to errors in estimating the values of the positions on the y-axis, but that the values on the x-axis are fixed. Such might be the case when test scores (which may have some error associated with their estimation) are the y-variable, and age (which can be determined exactly) is the x-variable. With points for an ROC curve there is error involved both in the determination of $z(S|s)$, the y-variable, and $z(S|n)$, the x-variable, so the method is inappropriate. There are least square methods for dealing with cases where both x- and y-variables have error associated with their estimation. In one such case, for example, it may be more appropriate to minimise, not the vertical distances of the points from the line, but the length of a perpendicular from each point to the line (see Figure 5.3 for an illustration of this). However, even this method requires that the points be estimated from independent data. The values of $z(S|s)$

and $z(S|n)$ in a rating scale task are all estimated from the same data and are not independent. No lease squares method is therefore appropriate for fitting ROC curves to points obtained from rating scale data and, until quite recently, experimenters were unable to do anything better than fit their curves by eye. This is not as bad as it seems as most rating scale experiments give points which leave little doubt about the path the ROC curve should take. Recently three papers (Dorfman & Alf, 1968, 1969; Ogilvie & Creelman, 1968) have presented a maximum likelihood ratio solution to the problem of fitting ROC curves. Their arguments are too complicated to present to the non-mathematical audience for whom this book is intended and the curve-fitting procedures they propose are best done by a computer programme. The researcher who wishes to fit curves by this technique would be advised either to persuade some-

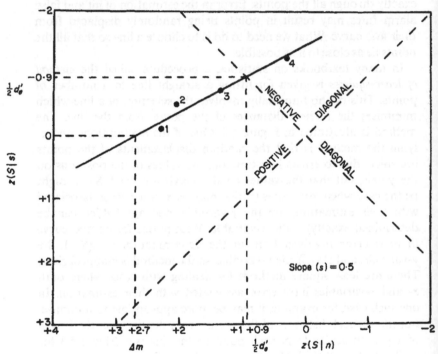

FIGURE 5.1 *A double-probability plot of the rating-scale data from Table 5.6. The* ROC *curve has been fitted by eye to the four points*

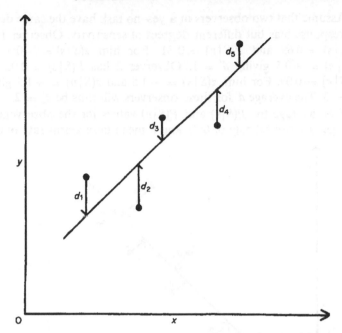

FIGURE 5.2   *Fitting a straight line to a set of points so that the vertical distances of the points (d_1, d_2, . . . d_5) are kept at a minimum. This procedure is appropriate when error is associated only with the variable on the y-axis*

one to write a programme for him, or to make use of existing programmes for analysing detection data. (See Appendix 4 for some of these.) In the interim, the examples used in this book, and many experimental data, are able to be fitted by eye without much difficulty. Such has been the method for determining the ROC curve passing through the points in Figure 5.1.

Averaging data over a number of individuals

In addition to obtaining points for the ROC curves for each observer, the experimenter may also wish to make a grand average of the points for the group. A method for doing this which reflects the central tendency of the group well as long as there are not large differences in the slopes of individual curves, has been proposed by Ingleby (1968). His procedure can be illustrated by a simple example.

111

Assume that two observers in a yes–no task have the same degree of response bias but different degrees of sensitivity. Observer 1 has $P(S|s) = 0.69$ and $P(S|n) = 0.31$. For him $z(S|s) = -0.5$ and $z(S|n) = +0.5$ giving $d' = 1$. Observer 2 has $P(S|s) = 0.93$ and $P(S|n) = 0.07$. For him, $z(S|s) = -1.5$ and $z(S|n) = +1.5$ giving $d' = 3$. The average d' for these observers will thus be $d' = 2$.

If we average the $P(S|s)$ and $P(S|n)$ values for the observers we will get a mean hit rate of 0.81 and a mean false alarm rate of 0.19.

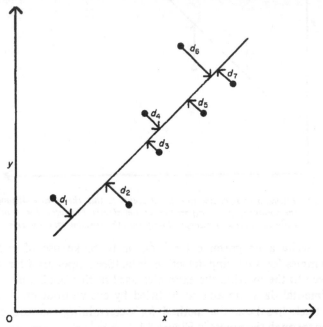

FIGURE 5.3 *Fitting a straight line to a set of points so that the perpendicular distances of the points from the line $(d_1, d_2, \ldots d_7)$ are kept at a minimum. This procedure is appropriate when error is associated with the determination of both x and y variables*

If these are used to calculate an average d' it will come out to be 1.76 which is incorrect. If we average $z(S|s)$ and $z(S|n)$ for the two observers we will get $z(S|s) = -1.0$ and $z(S|n) = +1.0$ which will give a mean $d' = 2$ which is correct.

In a rating scale task then, to find a set of points which will represent an average ROC curve for all the observers, one should average the $z(S \mid s)$ and $z(S \mid n)$ values for the observers for each category and not the P-values. By the same token the raw data for all the observers should not be combined and used to construct an overall ROC curve for the group.

The standard deviation of the signal distribution

The standard deviation of the signal distribution, σ_s, was defined in Chapter 4 as the reciprocal of the slope of the ROC curve. In Figure 5.1 it can be seen that as $z(S \mid n)$ increases by 1 z-unit, $z(S \mid s)$ increases by 0.5. The slope, s, is thus $0.5/1$ so that σ_s is $1/0.5 = 2$.

MEASURES OF OF SENSITIVITY

Δm and d'_e

As was seen in Chapter 4, Δm is the value of $z(S \mid n)$ at the point on the ROC curve where $z(S \mid s) = 0$. This occurs at the point x on the ROC curve in Figure 5.1 so that $\Delta m = 2.7$. d'_e is twice the value of either $z(S \mid s)$ or $z(S \mid n)$ at the point y where the ROC curve meets the negative diagonal. Thus $d'_e = 1.8$.

The measure $P(A)$

Another way of describing the observer's sensitivity would have been to have calculated $P(A)$, the proportion of the area under the ROC curve when plotted on scales of $P(S \mid s)$ and $P(S \mid n)$. This measure was first described in Chapter 2 where it was said that it could be found simply by plotting the curve on graph paper, joining the points by straight lines, and counting the squares underneath the curve. A second method for finding $P(A)$ will now be described. Its virtue is that it can be easily translated into a computer programme.

Figure 5.4 shows an ROC curve based on three points and plotted on scales of $P(S \mid s)$ and $P(S \mid n)$. The values of $P(S \mid s)$ and $P(S \mid n)$ are shown between the brackets beside each point with the $P(S \mid n)$ value first. Thus for point B, $P(S \mid n) = p$ and $P(S \mid s) = q$. The points for the ROC curve have been connected by straight lines.

It can be seen that the total area, $P(A)$, under the curve consists of four component areas; A_1, the area of the triangle ABF, and areas A_2, A_3 and A_4, the areas of the trapezia BCGF, CDHG, and

113

DEIH. A trapezium is a four-sided figure with two sides parallel, and its area is equal to half the distance between the parallel sides multiplied by the sum of the two parallel sides. A triangle can be thought of as a special case of a trapezium where one of the parallel sides has a length of zero, and of course its area is equal to half the length of its base multiplied by its altitude.

The coordinates of the points for the ROC curve allow us to find the

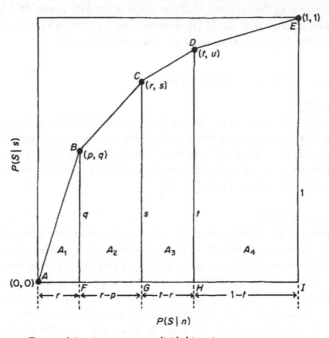

FIGURE 5.4　*An* ROC *curve divided into its component areas*

lengths of the sides necessary for calculating the four areas. These can be seen in Figure 5.4. Thus we can write

$P(A)$ (Total are under ROC curve)
$= A_1 + A_2 + A_3 + A_4$
$= \frac{1}{2}p(0+q) + \frac{1}{2}(r-p)(q+s) + \frac{1}{2}(t-r)(s+t) + \frac{1}{2}(1-t)(t+1).$

If, in general we have an ROC curve with N points, the general formula for $P(A)$, written in terms of $P(S|s)$ and $P(S|n)$, where $P_i(S|n)$ and

114

$P_i(S|s)$ are the coordinates for the ith point, is as follows:

$$P(A) = [P_1(S|n)-0][0+P_1(S|s)]$$
$$+ \tfrac{1}{2}[P_2(S|n)-P_1(S|n)][P_2(S|s)+P_1(S|s)] + \dots$$
$$+ \tfrac{1}{2}[P_i(S|n)-P_{i-1}(S|n)][P_i(S|s)+P_{i-1}(S|s)] + \dots$$
$$+ \tfrac{1}{2}[1-P_N(S|n)][1+P_N(S|s)]$$

$$= \tfrac{1}{2}\sum_{i=1}^{N+1}[P_i(S|n)-P_{i-1}(S|n)][P_i(S|s)+P_{i-1}(S|s)]. \quad (5.1)$$

In (5.1), $P_0(S|s)$ and $P_0(S|n)$ are the coordinates of the starting point of the ROC curve in the bottom left-hand corner of the graph and so are always equal to zero. $P_{N+1}(S|s)$ and $P_{N+1}(S|n)$ are the coordinates of the finishing point of the curve in the top right-hand corner and so are always equal to 1.

Using the geometric method we may now calculate the sensitivity measure $P(A)$ for the example using the $P(S|s)$ and $P(S|n)$ values in Table 5.4.

$$P(A) = \tfrac{1}{2}(0\cdot01-0\cdot00)(0\cdot00+0\cdot55)+\tfrac{1}{2}(0\cdot02-0\cdot01)(0\cdot55+0\cdot69)$$
$$+ \tfrac{1}{2}(0\cdot09-0\cdot02)(0\cdot69+0\cdot76)+\tfrac{1}{2}(0\cdot37-0\cdot09)(0\cdot76+0\cdot89)$$
$$+ \tfrac{1}{2}(1\cdot00-0\cdot37)(0\cdot89+1\cdot00) = 0\cdot886.$$

Sensitivity measures when some cells are empty

In Table 5.2 two variations of the basic example were considered involving empty cells in the raw data matrix. In case 1 the observer failed to make any false alarms in category 1, the strictest signal category. In case 2 the observer failed to make any hits in category 5, the strictest noise category. Both cases were dealt with by combining adjacent categories in Table 5.3 before calculating $P(S|s)$ and $P(S|n)$. The resulting $P(S|s)$ and $P(S|n)$ values appeared in Table 5.5.

What effect will this collapsing procedure have on the various sensitivity measures? Without a proper curve-fitting technique the effects of collapsing on Δm and d'_e cannot be assessed exactly. In general it can be said that the fewer the points for the ROC curve, the greater the difficulty in reliably fitting a straight line to the double-probability plot. Thus the necessity to collapse categories will result in less reliable estimates of Δm and d'_e. The extent of this unreliability will depend on how good and how numerous the remaining points are.

In the case of $P(A)$ we can be more definite about the effects of collapsing as it is possible to find $P(A)$ from the values of $P(S \mid s)$ and $P(S \mid n)$ for the two cases in Table 5.5. These $P(A)$ values can be compared with the one obtained from the uncollapsed category data.

Using (5.1) the following $P(A)$ values are found:

$$P(A) \text{ (case 1)} = \tfrac{1}{2}(0 \cdot 20 - 0 \cdot 00)(0 \cdot 00 + 0 \cdot 69) + \tfrac{1}{2}(0 \cdot 09 - 0 \cdot 02)(0 \cdot 69 + 0 \cdot 76) + \tfrac{1}{2}(0 \cdot 37 - 0 \cdot 09)(0 \cdot 76 + 0 \cdot 89) + \tfrac{1}{2}(1 \cdot 00 - 0 \cdot 37)(0 \cdot 89 + 1 \cdot 00) = 0 \cdot 884$$

$$P(A) \text{ (case 2)} = \tfrac{1}{2}(0 \cdot 01 - 0 \cdot 00)(0 \cdot 00 + 0 \cdot 55) + \tfrac{1}{2}(0 \cdot 02 - 0 \cdot 01)(0 \cdot 55 + 0 \cdot 69) + \tfrac{1}{2}(0 \cdot 09 - 0 \cdot 02)(0 \cdot 69 + 0 \cdot 76) + \tfrac{1}{2}(1 \cdot 00 - 0 \cdot 09)(0 \cdot 76 + 1 \cdot 00) = 0 \cdot 861.$$

In both cases $P(A)$ is underestimated compared with the value of $0 \cdot 886$ obtained from the uncollapsed category data. The reason for this underestimation is illustrated in Figure 5.5. However it is heartening to notice that the error in $P(A)$ due to collapsing is not large. In fact the $P(A)$ measure can be taken from data too poor for determining Δm or d'_e. The reason for $P(A)$'s tolerance of bad data, as opposed to that of the other two measures, is that curve fitting does not enter into $P(A)$'s determination. The ROC curve points are plotted out, joined by straight lines, and the area calculated. This offers some hope to the experimenter, who through force of circumstances is unable to conduct the large number of signal and noise trials necessary to determine the path of the ROC curve on the double-probability scale and, hence the values of Δm or d'_e. It is not easy to say how few signal and noise trials an experimenter can get away with if he decides to opt for $P(A)$ rather than Δm or d'_e as his sensitivity measure. As in any experiment, the answer to this question depends on how strong the experimental treatments are relative to uncontrolled sources of error. As was mentioned earlier some success has been obtained with as few as fifty signal and fifty noise trials but it is worth noting that in that case experimental error due to individual differences was minimized by having all observers work under all conditions.

Transforming P(A) values before statistical analysis

Having measured the sensitivity of groups of observers under several conditions, the experimenter will wish to assess statistically the effects of his experimental treatments. Perhaps his experiment involves some analysis of variance design. The sensitivity measures, whether they be Δm, d'_e or $P(A)$ can be treated in the same way as any performance scores for the purposes of statistical analysis. If the

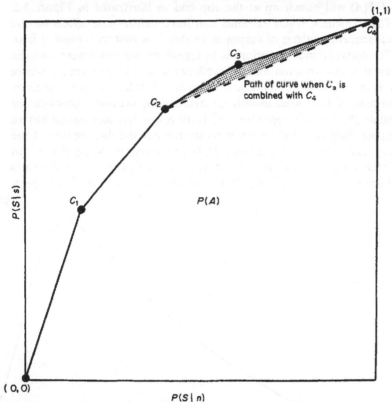

FIGURE 5.5 *Effect on P(A) of collapsing rating categories. When category C_3 is combined with C_4 the curve follows the dotted line, and the shaded area is omitted from P(A)*

experimenter has been fortunate enough to obtain either Δm or d'_e for each observer in each experimental condition, there are probably few problems in conducting his statistical tests. Both these

117

scores, being z-scores, can range from $-\infty$ to $+\infty$ although negative values will be rarely found. With luck they will be normally distributed, or sufficiently so for an analysis of variance to be carried out.

However, if $P(A)$ is used as a sensitivity score, problems can arise. As this measure is a probability score it has an upper limit of 1. If some treatment conditions yield high sensitivity, the distribution of $P(A)$ will bunch up at the top end as illustrated in Figure 5.6. This skewness, if too extreme, can have unfortunate effects on the subsequent analysis of variance, so that it is best to remove it first. The conventional procedure is to transform the raw scores, and the usual transformation for probabilities is to take $2 \arcsin \sqrt{p}$ where p is the raw probability. Using $2 \arcsin \sqrt{P(A)}$ for statistical analysis instead of $P(A)$ itself usually overcomes the skewness problem for while $P(A)$ has an upper limit of 1, the arcsin transformation has an upper limit of π and so tends to stretch out the distribution at the top end. Tables of $2 \arcsin \sqrt{P(A)}$ are given in Appendix 5. An interesting feature of this transformation of $P(A)$ scores in shown in Figure 5.7. Here $2 \arcsin \sqrt{P(A)}$ is plotted against d'. As can be seen,

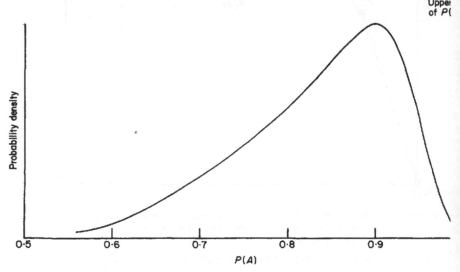

FIGURE 5.6   *The tendency of P(A) values representing high levels of sensitivity to bunch at the upper limit of 1·0 thus producing a skewed distribution*

118

the result is a negatively accelerating function but the curve is very close to linear up to about $d' = 3$. Thus, over quite a wide range of values, $2 \arcsin \sqrt{P(A)}$ can be treated as an approximation to d'.

MEASURES OF BIAS

The measures of x and β

The bias represented by each criterion point on the rating scale can be described either in terms of x, the distance of a criterion from the mean of the noise distribution or in terms of β the ratio of the ordinates of the two distributions at each criterion. No further calculations are needed to find x as it is the value of $z(S \mid n)$ for each criterion and for the example, these values have been given in Table 5.6.

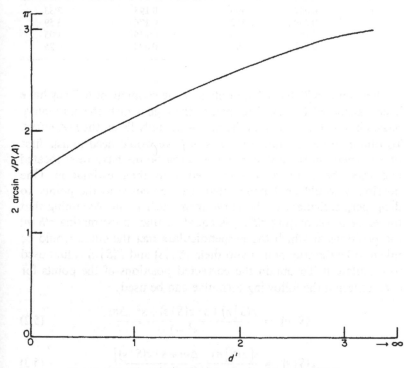

FIGURE 5.7   *The relationship between d' and $2 \arcsin \sqrt{P(A)}$*

To find β, y_n is first found for each $P(S \mid n)$ value by using the normal curve tables of Appendix 5. Next, the same tables are used to find the signal ordinate for each $P(S \mid s)$ value. As the standard deviation of the signal distribution has been found to be equal to 2, each ordinate must be divided by this standard deviation to obtain y_s, the height of the signal distribution at each criterion. Finally, β for each criterion is found by dividing each y_s value by its corresponding y_n value. These calculations are shown in Table 5.7

TABLE 5.7   *Calculation of β values for the example*

Criterion	y_n	Signal ordinate	$y_s = \dfrac{\text{Signal ord.}}{s}$	$\beta = \dfrac{y_s}{y_n}$
1	0·027	0·396	0·198	7·33
2	0·049	0·352	0·176	3·59
3	0·163	0·310	0·155	1·05
4	0·378	0·187	0·094	0·25

The values in Table 5.7 are only a crude estimate of β. They have been calculated for the four points through which the ROC curve passes but, in fact, not all of the points actually lie on the ROC curve. To obtain more accurate estimates of β we would need to ask, first of all, where on the ROC curve would the points have been located had there been no error associated with their estimation. One possibility would be, having fitted the ROC curve to the points, to drop perpendiculars to the curve from each point. Assuming that the error in estimating $z(S \mid s)$ is equal to that in estimating $z(S \mid n)$ the positions at which the perpendiculars met the curve would be taken to be the true points and their $P(S \mid s)$ and $P(S \mid n)$ values used to calculate β. To obtain the corrected positions of the points for each criterion the following formulae can be used:

$$z(S \mid n)' = \frac{z(S \mid n) + s \cdot z(S \mid s) + s^2 \cdot \Delta m}{s^2 + 1} \tag{5.2}$$

$$z(S \mid s)' = \frac{s[z(S \mid n) - \Delta m + s \cdot z(S \mid s)]}{s^2 + 1}, \tag{5.3}$$

where

$z(S \mid s)$ is the z-score corresponding to the observed hit rate of the criterion,

$z(S \mid n)$ is the z-score corresponding to the observed false alarm rate of the criterion,

s is the slope of the ROC curve,

$z(S \mid s)'$ is the corrected value of $z(S \mid s)$ for a perpendicular dropped to the ROC curve and,

$z(S \mid n)'$ is the corrected value of $z(S \mid n)$ for a perpendicular dropped to the ROC curve.

In the present example it is doubtful that the use of (5.2) and (5.3) would give better estimates of β than using the uncorrected positions of the points. For one thing, all four points lie close to the line. For another, the curve was fitted by eye which introduces another source of error and also provides no information as to whether the error of estimation of $z(S \mid s)$ and $z(S \mid n)$ was in fact equal.

An experimenter who wants to know whether an experimental treatment has produced changes in response bias would be well advised to look at the obtained values of $z(S \mid n)$ (or $P(S \mid n)$ for that matter) in addition to calculating β. Consider the following hypothetical experiment where d' and β are measured under two conditions. Figure 5.8 shows the changes produced in the signal and noise distributions.

It can be seen that d' for treatment B is larger that that for treatment A, and that in terms of β the criterion in treatment B is more risky than that in treatment A. By looking at the β values we might have concluded that the two conditions have different effects on the criterion. In fact, what has happened is that the observer has kept his criterion at a fixed distance, x_c, from the noise distribution mean rather than adjusting it to maintain a constant β under both conditions. Has the criterion really changed in this case?

Statistical analysis of x and β scores

Suppose that an experiment has used a five-point rating scale to obtain ROC curves from a group of ten observers, each observer

121

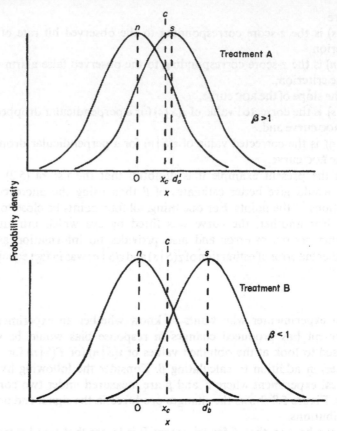

FIGURE 5.8 *A criterion, C, which is held at a constant distance, x_c, from the noise distribution mean but whose β value changes as d' changes*

working under two conditions and thus yielding two curves. Subsequently the experimenter wishes to examine statistically, the differences in either β or x under two conditions. At first glance, a three-way analysis of variance design might seem appropriate, the three factors being experimental conditions (two levels), rating categories (four levels), and observers (ten levels), with β (or x) being measured under each combination of levels of these factors. This, in fact, is not appropriate. The reason is the β or x values for each rating category are based on the same raw data and are not

122

independent. The best that can be done, it appears, is to test the difference in bias between the two conditions separately for each rating category. In this case, with two experimental conditions, it would be more appropriate to do a t-test between experimental conditions for each rating category, a total of four such tests, rather than a single $2 \times 4 \times 10$ factorial analysis of variance.

If separate yes–no tasks had been used to determine the bias measures for each category, then the analysis of variance is possible as different data have been used to estimate each β and x.

A practical point to remember before conducting statistical tests on β values is that β has a lower limit of 0. It cannot assume negative values. This may mean that the distribution of β values from an experimental condition may be skewed, and if the statistical test to be used requires a roughly normal distribution of scores, the βs will need to be transformed. In Chapter 3 the transformation $\log \beta$ was mentioned in connection with graphical presentation of results. This transformation will normally prove useful in removing skew due to the scores bunching up near 0.

Non-parametric assessment of response bias

If an experimenter doubts that his obtained hit and false alarm rates are sufficiently good to calculate Δm and d'_e then he should have even less confidence in being able to obtain reliable β values for the criteria. At least the sensitivity measures are based on a line fitted to all the points, but in determining β each point on the ROC curve must stand on its own merits. If a criterion is placed well into the tails of the signal and noise distributions, quite small errors in the estimation of hits or false alarms can cause large errors in the estimation of the heights of the distributions at the criterion. Nor is there a satisfactory non-parametric alternative to β in the same way that $P(A)$ can be used in place of Δm and d'_e.

Rather than giving up all hope of measuring bias when there is insufficient data to obtain β values for each criterion, a method will be proposed which enables the experimenter to make an overall assessment of bias. The method has its limitations; it provides only a single bias score and not one for each point, but this is all that many experimenters require.

The aim of the method is to find the point B on the observer's rating scale at which he is equally disposed to signal and noise responses.

Figure 5.9 shows the distributions of signal and noise for two hypothetical observers for a yes–no rating task. Signal and noise distributions have equal variances and both observers have $d' = 1$. It can be seen that the observers differ in the placement of their five criteria, observer 1's criteria being located further up the x-axis than the criteria of observer 2. Table 5.8 shows the $P(S|s)$ and

TABLE 5.8 $P(S|s)$, $P(S|n)$, $P(S|s)+P(S|n)$ and β for the criteria of the two observers whose signal and noise distributions are shown in Figure 5.9

Observer 1

Observer's response
High certainty signal to high certainty noise

Category	1	2	3	4	5
$P(S|s)$	0·16	0·31	0·50	0·69	0·84
$P(S|n)$	0·02	0·07	0·16	0·31	0·50
$P(S|s)+P(S|n)$	0·18	0·38	0·66	1·00	1·34
β	4·8	2·7	1·7	1·0	0·6

Observer 2

Observer's response
High certainty signal to high certainty noise

Category	1	2	3	4	5
$P(S|s)$	0·50	0·69	0·84	0·93	0·98
$P(S|n)$	0·16	0·31	0·50	0·69	0·84
$P(S|s)+P(S|n)$	0·66	1·00	1·34	1·52	1·82
β	1·7	1·0	0·6	0·4	0·2

$P(S|n)$ values for each criterion for the two observers. Values of β for each criterion have been calculated and it can be seen that those for observer 1 are greater than those for observer 2.

Also shown in Table 5.8 is the sum $P(S|s)+P(S|n)$ for each

124

criterion. This sum is twice the proportion of S responses made at each criterion point. In the equal variance case, at the criterion where $\beta = 1$ (category 4 for observer 1 and category 2 for observer 2) the observer will make signal and noise responses with equal probabilities so that $P(S|s) + P(S|n)$ for this criterion will be equal to 1.

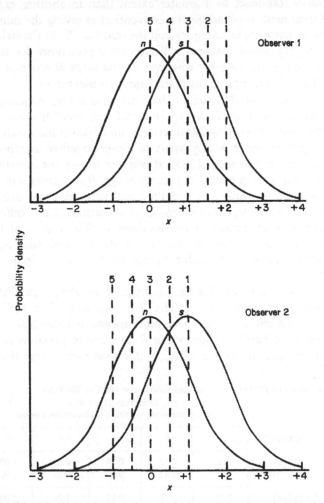

FIGURE 5.9   *Placement of criteria for two observers with equal sensitivity*

125

B, the non-parametric measure of response bias can thus be defined as the rating scale category at which $P(S|s) + P(S|n) = 1$.

For observer 1, $B = 4$ and for observer 2, $B = 2$.

The measure B may be useful when an experimenter wishes to know whether one experimental treatment has caused his observers to favour S responses to a greater extent than in another experimental treatment. If we adopt the convention of giving the number '1' to the strictest signal category and the number 'N' to the strictest noise category, small B scores will indicate a preference for signal responses (as in the case for observer 2) and large B scores a preference for noise responses (as is the case for observer 1).

One limitation of B is that it does not give a bias measure for each criterion, so that if an experimental treatment has resulted, not in the observer moving all his criteria up or down the x-axis, but in spacing them more widely apart or closer together as reported by Broadbent & Gregory (1963), B will not detect these changes. Another point to remember is that although B coincides with the rating category at which $\beta = 1$ in the equal variance case, such will not happen when signal and noise distributions have different variances. The advantage of the measure is that it does at least allow a crude assessment of bias when the hit and false alarm probabilities are too unreliable to obtain β or x for individual criteria.

B will now be obtained for the example whose $P(S|s)$ and $P(S|n)$ values are given in Table 5.4. The first step is to find $P(S|s) + P(S|n)$ for each category and these sums appear in Table 5.9.

Inspection of Table 5.9 shows that B does not lie precisely at any category. It must lie above category 3 whose sum is less than 1,

TABLE 5.9 $P(S|s) + P(S|n)$ *for the hit and false alarm rates of Table* 5.4

Observer's response

High certainty signal to high certainty noise

Category	1	2	3	4	5
$P(S|s)$	0·55	0·69	0·76	0·89	1·00
$P(S|n)$	0·01	0·02	0·09	0·37	1·00
$P(S|s) + P(S|n)$	0·56	0·71	0·85	1·26	2·00

126

but below category 4 whose sum exceeds 1. We must therefore find *B* by interpolation between these two categories. The procedure is quite like that for the calculation of a median from frequency data when the median lies somewhere between two adjacent frequency classes. We proceed as follows:

(a) The categories between which *B* must lie are determined. Category 3 is the lower category and category 4, the upper category.

(b) The number of $[P(S|s)+P(S|n)]$ units between the upper and lower categories is $1·26 - 0·85 = 0·41$.

(c) As *B* is the point where $P(S|s)+P(S|n) = 1$, then it lies $1 - 0·85 = 0·15 \; [P(S|s)+P(S|n)]$ units above the lower category.

(d) Assuming that the $0·41 \; [P(S|s)+P(S|n)]$ units between the two categories are equally spaced, then *B* must occupy a position $0·15/0·41$ of the way between the lower and upper categories.

(e) In category units, the distance between the lower and upper categories is $4 - 3 = 1$ unit. As *B* lies $0·15/0·41$ of the way into this unit, it will be $(0·15/0·41) \times 1$ category units above category 3.

(f) Thus *B* will be equal to the category unit value of the lower category, plus the distance (in category units) it lies above the lower category, namely, $3 + 0·37 = 3·37$.

These steps can be expressed in a single formula.

$$B = \frac{1 - P_l(S|s) - P_l(S|n)}{P_u(S|s) + P_u(S|n) - P_l(S|s) - P_l(S|n)} \cdot S + C_l \qquad (5.4)$$

where
$P_l(S|s)$ is the hit probability for the lower category,
$P_l(S|n)$ is the false alarm probability for the lower category,
$P_u(S|s)$ is the hit probability for the upper category,
$P_u(S|n)$ is the false alarm probability for the upper category,

S is the distance between the upper and lower categories in category units (normally it will be 1 and can thus be dropped from the formula), and
C_l is the size of the lower category in category units.

Estimating B when end categories have been collapsed or when B falls in an end category

It was seen earlier that observers may fail to make hits or false alarms in the extreme categories of the rating scale with the consequence that some $P(S|s)$ values· may turn out to be 1 and some $P(S|n)$ values may turn out to be 0. To obtain sensitivity scores, such offending categories were combined with adjacent ones. When it comes to finding B however, it is better to revert to the uncollapsed data. To take case 2 of Table 5.2 as an example, the hit and false alarm rates for the raw data without collapsing categories are as given in Table 5.10. By substituting in (5.4), B is found to be 3·29.

TABLE 5.10 $P(S|s)$, $P(S|n)$ and $P(S|s) + P(S|n)$ for the raw data of Table 5.2

	Observer's response High certainty signal to high certainty noise				
Category	1	2	3	4	5
$P(S\|s)$	0·55	0·69	0·76	1·00	1·00
$P(S\|n)$	0·01	0·02	0·09	0·37	1·00
$P(S\|s) + P(S\|n)$	0·56	0·71	0·85	1·37	2·00

A more serious problem occurs when either the value of $P(S|s) + P(S|n)$ is greater than 1 in the first category of the rating scale, or less than 1 for the last category. In the first case B lies below category 1 so while category 1 is the upper category, no lower category exists. A crude solution is to assume the existence of a category 0 which has $P_l(S|s) = P_l(S|n) = 0$. Likewise in the second case B lies above category N, the last category on the rating scale so the category N becomes the lower category. Here we could assume the existence of category $(N+1)$ as the upper category with $P_u(S|s) = P_u(S|n) = 1$.

128

Neither of these methods is likely to give a good estimate of B but they provide a way out for the experimenter who needs bias scores for the purpose of statistical analysis but encounters one or two observers in his sample with extreme biases to signal or noise.

Problems

1. From the raw data given below find, for each observer:
(a) Δm,   (b) d'_e (c) $P(A)$.

		Observer's response High certainty signal to high certainty noise				
Category		1	2	3	4	5
Observer 1	s	34	28	14	20	4
	n	16	11	7	24	42
Observer 2	s	42	20	17	9	12
	n	0	21	25	23	31
Observer 3	s	38	16	19	27	0
	n	10	14	34	24	18

2. The following hit and false alarm rates are obtained from four observers. Find $P(A)$ and B for each observer.

		Observer's response High certainty signal to high certainty noise			
Category		1	2	3	4
Observer 1	$P(S\mid s)$	0·38	0·52	0·80	0·94
	$P(S\mid n)$	0·02	0·06	0·16	0·38

129

Observer 2	$P(S\,|\,s)$	0·64	0·88	0·97	0·98
	$P(S\,|\,n)$	0·08	0·24	0·50	0·60
Observer 3	$P(S\,|\,s)$	0·20	0·36	0·66	0·78
	$P(S\,|\,n)$	0·04	0·10	0·28	0·52
Observer 4	$P(S\,|\,s)$	0·48	0·78	0·90	0·94
	$P(S\,|\,n)$	0·16	0·42	0·66	0·78

Chapter 6

CHOICE THEORY APPROXIMATIONS TO SIGNAL DETECTION THEORY

Signal detection theory is not the only theory concerned with the measurement of sensitivity and response bias. Luce (1959, 1963) has presented his choice theory, which resembles detection theory in many ways. There is some controversy as to which of these two points of view is the most appropriate for application to psychological problems, but we will not be concerned with this matter here. The fact is that choice theory and detection theory have marked similarities, so much so that in some cases the one can be used as an approximation to the other. Choice theory can be applied to some types of experimental task which detection theory cannot, and provides another useful tool for the experimenter interested in sensitivity and bias problems. This close resemblance between the two theories has been exploited by Broadbent (1967), Broadbent and Gregory (1967) and Ingleby (1968). This chapter is concerned with explaining Broadbent's and Ingleby's methods of using choice theory as a way of gathering information about sensitivity and bias in experimental situations which do not involve the use of recognition tests.

In the previous chapters, either no assumptions have been made about the nature of the underlying distribution of signal and noise, or it has been assumed that they were Gaussian. In this chapter we will consider another distribution called the *logistic distribution* which closely resembles the normal one. The similarities between the two distributions are close enough for the logistic distribution to be able to be used as an approximation to the normal. Why should we wish to use such an approximation?

First, the normal distribution is difficult to work with. Formulae

(3.5(a)) and (3.5(b)) are formulae for a standard normal distribution. They enabled the height of the distribution to be determined at various distances from its mean. The basic formula, as you will recall, was

$$y = \frac{e^{-\frac{1}{2}x^2}}{\sqrt{(2\pi)}},$$

where y was the height of the distribution and x, the distance from the distribution mean.

Not only have we needed to determine y at various distances from the distribution mean but we have also wished to find the area under the curve which lies to one or other side of a given x value. This has always been done by looking up the appropriate P value in normal-curve area tables, but how were these composed in the first place? The answer is, by a mathematical technique called integral calculus. The area above x is given by the formula

$$A = \int_{x}^{+\infty} y\, dz. \tag{6.1}$$

For those who are unfamiliar with the calculus, the right-hand side of (6.1) can be interpreted in the following manner. Imagine that the area lying above x has been divided into a number of very narrow strips, each of these having a base dz units long. This is illustrated in Figure 6.1. The figure also shows that each strip will have a height of $y_1, y_2, y_3, \ldots$ etc. Thus the area of each strip will be approximately equal to that of a rectangle with base dz and height y. That is, each area will be $y\, dz$. The integral sign in the formula, $\int$, is analogous to the familiar summation sign, $\sum$. It tells us to start at x and add together all the values of $y\, dz$ until we reach $+\infty$. This procedure is easy to understand in principle, but in practice the determination of areas under a normal curve by integration is very tedious. This, of course, is, why normal curve tables have been prepared. It will be seen presently that the logistic distribution curve is very easily integrated.

The more tractable nature of the logistic distribution function has resulted in it being used for fitting ROC curves to sets of $z(S\,|\,s)$ and $z(S\,|\,n)$ scores. In Chapter 5 it was mentioned that the least-

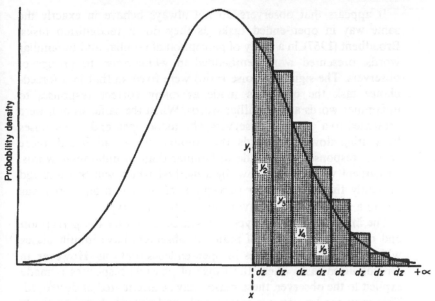

FIGURE 6.1   *Illustration of the determination of the area under the curve to the right of x from the integral $\int_x^{+\infty} y\,dz$*

squares technique is inappropriate and that Ogilvie and Creelman (1968) have devised another method for curve fitting in this case. Their technique involves, in part, the use of the logistic distribution in place of the normal one.

Second, the logistic distribution is the distribution which is basic to the development of Luce's choice theory. This is why choice theory and detection theory, both developed from very different axioms, resemble one another so closely in many cases. Some psychological phenomena are not very amenable to investigation by detection theory methods. The methods that have been considered have been those of the yes–no, rating scale and forced-choice tasks. All of these are recognition tests where the experimenter makes explicit to the observer the set of possibilities from which he must select his response. On the other hand, psychologists have also used recall tasks to assess memory and perception, and in these situations there is no well-defined set of possibilities given to the observer. He is quite free to give any response he chooses. Tasks of this type could be called *open-ended* ones.

133

It appears that observers do not always behave in exactly the same way in open-ended tasks as they do in recognition tasks. Broadbent (1967), in a study of perception of familiar and unfamiliar words, presented words embedded in white noise to groups of observers. The signal-to-noise ratios were fixed so that in a forced-choice task the observers made as many correct responses to unfamiliar words as to familiar words. When the same stimuli were presented to a group of observers who made open-ended responses by writing down the words they thought they had heard, more correct responses were made to familiar than to unfamiliar words. Broadbent was able to show, by a method which will be discussed presently that the superior perception of the common words was due to a response bias in favour of words of this type.

The implication of this type of research is that for the perception and recall of some types of material, observers have inbuilt biases which affect their responses in open-ended situations. However, in a recognition task, where the range of possible responses is made explicit to the observer, these biases may be attenuated or destroyed. This may render yes–no, rating scale and forced-choice methods unsuitable for collecting information about criterion changes. Choice theory, on the other hand, can be used to detect shifts in a criterion from data of an open-ended type. Before we can discuss how open-ended data can be used to find measures analogous to d' and β we will have to talk a little more about the logistic distribution.

THE LOGISTIC DISTRIBUTION

The simplified formula for the height, y, of the logistic distribution at a distance z standard deviations from its mean is

$$y = \frac{e^z}{(1+e^z)^2},$$ (6.2)

where e is the mathematical constant introduced in Chapter 3 and whose value is approximately $2\cdot718$.

Formula (6.2) can be used as an approximation to formula (3.5), the height of a normal distribution curve. In (6.2) the cri-

134

terion, C, has been set at a distance of z standard deviations from the mean. We wish to find the areas under the curve to the right and to the left of C. From the previous discussion in this chapter this can be done by integrating (6.2). The formula for the integral is simple and will be stated here without any derivation. For those interested in the derivation, it is given in Appendix 3.

If C lies *below* the mean of the distribution, the proportion of the area *above* C is

$$P_1 = \frac{e^z}{1+e^z}. \tag{6.3(a)}$$

As the total area under the curve is equal to 1, then the area *below* C is 1 minus (6.3a), which is:

$$P_2 = \frac{1}{1+e^z}. \tag{6.3(b)}$$

If C lies *above* the mean then the area *above* C is given by (6.3(b)), and the area below C by (6.3(a)). Thus (6.3(a)) gives the larger of the two areas and (6.3(b)), the smaller of the two areas.

We will now take a step, the usefulness of which will become apparent shortly. The x-axis on which the distribution is scaled has been calibrated in units of z, the standard deviation. Any value of z can be thought of as the logarithm of some other number x so that in place of z in (6.3(a)) and (6.3(b)) we could write $\log x$. Thus in these formulae, the term e^z can be replaced by $e^{\log x}$ where $\log x = z$. In Chapter 3 it was seen that $e^{\ln x}$ was equal to x, so that if we ignore the fact that we are working in logarithms to the base 10 and not to the base e, we can write x in place of $e^{\log x}$ in these formulae. Hence, if C is $\log x$ standard deviations *below* the mean of the distribution, (6.3(a)) becomes

$$P_1 = \frac{x}{1+x}, \tag{6.4(a)}$$

and (6.3(b)) becomes

$$P_2 = \frac{1}{1+x}. \tag{6.4(b)}$$

These areas are illustrated in Figure 6.2. Of course, if C lies above the mean, (6.4(b)) will give the area above C, and (6.4(a)), the area below C.

135

K

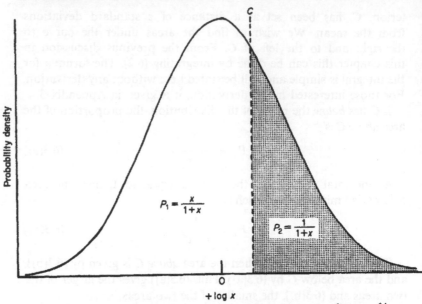

Probability density

$$P_1 = \frac{x}{1+x}$$

$$P_2 = \frac{1}{1+x}$$

+ log x

FIGURE 6.2   *Areas to the left and the right of a criterion, C, set at a distance of log x from the distribution mean. Formulae for areas are based on the approximation of the logistic to the normal distribution when the x-axis is scaled in units of log x*

DETERMINING DETECTION MEASURES FROM LOGISTIC DISTRIBUTIONS

Figure 6.3 shows a signal and a noise distribution drawn on an x-axis scaled in units of log x. The reference point of zero from which all distances along the x-axis are measured is midway between the distribution means. This means that the noise distribution mean is $-\log\alpha$ from this point and the signal distribution mean $+\log\alpha$ from 0. The observer's criterion, C, is, in this case, $+\log v$ units above 0. It can be seen that d', the distance between the means of the distribution, is equal to $2\log\alpha$. It is assumed in all of the following discussion that we are dealing with distributions of signal and noise with equal variances.

Consider the noise distribution first. With the criterion set at $+\log v$ from 0, what are the values of $P(S|s)$ and $P(S|n)$? $P(S|n)$ will be the proportion of the noise distribution which lies above C. C lies $\log\alpha + \log v$ standard deviations above the noise distribu-

136

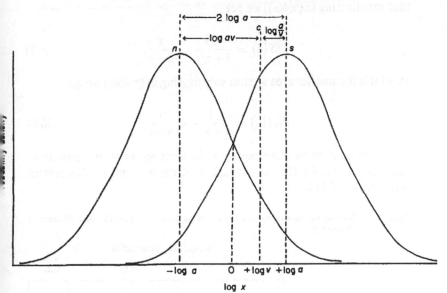

FIGURE 6.3   *Distributions of signal and noise for Luce's choice theory.* $d' = 2 \log \alpha$, *and the criterion, C, is set at* $+ \log v$

tion mean. In terms of (6.4), $\log x = \log \alpha + \log v$ so that $x = \alpha v$. As C lies above the mean of the noise distribution, $P(S \mid n)$ will be the smaller area and will be given by (6.4(b)) so that

$$P(S \mid n) = \frac{1}{1 + \alpha v}. \tag{6.5}$$

$P(N \mid n)$ will be the larger area and therefore given by (6.4(a)), so that

$$P(N \mid n) = \frac{\alpha v}{1 + \alpha v}. \tag{6.6}$$

Looking next at the signal distribution it is seen that C lies $\log \alpha - \log v$ standard deviations below the signal distribution mean. Again in terms of (6.4), $\log x = \log \alpha - \log v$ so that $x = \alpha / v$. As C lies below the signal distribution mean, $P(S \mid s)$ is the larger area so

137

that substituting in (6.4(a)) we get

$$P(S\mid s) = \frac{\alpha/v}{1+\alpha/v} = \frac{\alpha}{\alpha+v}. \tag{6.7}$$

$P(N\mid s)$ is the smaller area so that substituting in (6.4(b)) we get:

$$P(N\mid s) = \frac{1}{1+\alpha/v} = \frac{v}{\alpha+v}. \tag{6.8}$$

Formulae (6.5) to (6.8) can now be used to write the stimulus-response matrix for the yes–no task in terms of α and v. This matrix is shown in Table 6.1.

TABLE 6.1 *Stimulus-response matrix for a yes–no task based on the distributions of Figure 6.3*

| | | Response alternative | | Row total |
		S	N	
Stimulus alternative	s	$P(S\mid s) = \dfrac{\alpha}{\alpha+v}$	$P(N\mid s) = \dfrac{v}{\alpha+v}$	1
	n	$P(S\mid n) = \dfrac{1}{1+\alpha v}$	$P(N\mid n) = \dfrac{\alpha v}{1+\alpha v}$	1

It can be seen that the two variables, α and v, of which the probabilities in Table 6.1 are composed, are measures of sensitivity and bias. The variable α is a sensitivity component, as in Figure 6.3 the distance between the distribution means was $2\log\alpha$. The variable v is a measure of bias, as in Figure 6.3 the observer's criterion was set at $+\log v$. Given a set of such probabilities obtained from a yes–no task, these two measures can be obtained by some simple multiplication.

Starting with α,

$$\frac{P(S\mid s)\cdot P(N\mid n)}{P(S\mid n)\cdot P(N\mid s)} = \left(\frac{\alpha}{\alpha+v}\cdot\frac{\alpha v}{1+\alpha}\right)\Big/\left(\frac{1}{1+\alpha v}\cdot\frac{v}{\alpha+v}\right)$$

$$= \alpha^2.$$

As $d' = 2 \log \alpha$,

$$d' = \log\left(\frac{P(S \mid s) \cdot P(N \mid n)}{P(S \mid n) \cdot P(N \mid s)}\right). \tag{6.9}$$

Next, for v,

$$\frac{P(N \mid s) \cdot P(N \mid n)}{P(S \mid s) \cdot P(S \mid n)} = \left(\frac{v}{\alpha+v} \cdot \frac{\alpha v}{1+\alpha v}\right) \bigg/ \left(\frac{\alpha}{\alpha+v} \cdot \frac{1}{1+\alpha v}\right)$$

$$= v^2.$$

So that $\log v$, the position of the observer's criterion on the x-axis is

$$\log v = \log\left(\frac{P(N \mid s) \cdot P(N \mid n)}{P(S \mid s) \cdot P(S \mid n)}\right)^{\frac{1}{2}}. \tag{6.10}$$

Formulae (6.9) and (6.10) allow the computing of sensitivity and bias measures for a yes–no task directly from the stimulus-response matrix and without resort to normal-curve tables.

Furthermore, it is possible, by replacing the normal distribution by the logistic one, to obtain α and v for 2AFC and mAFC tasks. All of this appears in Luce (1959, 1963). For yes–no and 2AFC tasks the choice theory approximations to the measures obtained by detection theory are very close. Ingleby (1968) has shown however, that in the mAFC task the relationship between α and the number of alternatives from which the observer must select his response is not quite the same as the relationship between d' and number of alternatives. However he was unable to find any published mAFC data which supported the one theory in preference to the other.

THE MATRIX OF RELATIVE RESPONSE STRENGTHS

Before proceeding to the analysis of open-ended tasks there is one more step to be taken. If the denominators of the expressions in Table 6.1 are removed, Table 6.2 is obtained. This table is called a matrix of *relative response strengths*. This matrix may be interpreted as follows:

139

TABLE 6.2   *The matrix of relative response strengths for the yes–no task*

		Response alternative S	N
	s	α	v
Stimulus alternative	n	1	αv

The ratio of the pair of entries within a particular row is equal to the ratio of the probabilities of the two responses.

This may be made clearer by an example.

Consider that event s has occurred, that is, we are dealing with the top row of Table 6.2. From the definition, the ratio of the two entries, α and v, in the row is equal to the ratio of the probability of responding S to the probability of responding N given that s occurred. In other words $\alpha/v = P(S\,|\,s)/P(N\,|\,s)$.

Also, had event n occurred the definition implies that $1/\alpha v = P(S\,|\,n)/P(N\,|\,n)$. You should be able to see that these expressions are correct by looking back at Table 6.1.

In writing matrices of relative response strengths there is no need to restrict the situation to just two stimuli and two responses. Table 6.3 shows a matrix of relative response strengths for a 3AFC task. Each stimulus (S_1, S_2, S_3) has a corresponding response (R_1, R_2, R_3). Looking first at the columns of the matrix it can be seen that each response has a bias on it represented by the terms v_1, v_2 and v_3. These response biases are independent of any stimulus effect so that no matter if stimulus S_1, S_2, or S_3 is presented, response R_1, for example, will always have a response strength of at least v_1.

TABLE 6.3   *The matrix of relative response strengths for a 3AFC task*

		Response alternative 1	2	3
	1	$\alpha_1 v_1$	v_2	v_3
Stimulus alternative	2	v_1	$\alpha_2 v_2$	v_3
	3	v_1	v_2	$\alpha_3 v_3$

Looking next along the diagonal of the matrix, the correct response to each stimulus has a sensitivity component denoted by the αs. Each of these may be different, meaning that d' for some stimuli may be greater than d' for others.

Again for this matrix the ratio of a pair of entries within a particular row is equal to the ratio of the probabilities of occurrence of the two responses. If $P(R_1 | S_2)$ is the probability of giving response R_1 when stimulus S_2 was presented, and $P(R_2 | S_2)$ the probability of giving response R_2 to stimulus S_2, then by definition; $P(R_2 | S_2)/P(R_1 | S_2) = \alpha_2 v_2/v_1$.

OPEN-ENDED TASKS

From the information in the preceding sections it is now possible to explain Broadbent's method for measuring sensitivity and bias in an open-ended task.

Consider that there are two classes of stimuli and that we are interested in seeing whether items of one class are perceived better than items of the other. To make the example more explicit, assume that the stimuli are words. There are many ways in which words might be divided into two classes; common or uncommon, emotional or neutral, abstract or concrete, etc. In the example we will refer to the stimuli as being either class A or class B words. Assume that we have found out by some means or another that there are, in the language, a total of N_A words of class A and N_B words of class B. Using Table 6.3 as a guide it is possible to write a matrix of relative response strengths for all the words in the two classes. Altogether the matrix will have $N_A + N_B$ rows and columns. To simplify things it will be assumed that, on the average, all words from class A are equally discriminable, so that in choice theory terms the sensitivity for each class A word will be equal to α_A. Similarly sensitivity to each class B word will be α_B. We wish to determine experimentally whether α_A differs from α_B.

Perhaps it is also suspected that observers are more disposed to giving words of one class as responses more often than words of the other class. Broadbent (1967) found that observers gave responses which were common words more often than those which were uncommon words, no matter whether the stimulus itself had been

141

a common or an uncommon word. So, in the matrix of relative response strengths, a response bias component must be added. Let us arbitrarily put a bias of v on each class A word and set $v = 1$ for the class B words. The matrix of relative response strengths can now be written. It is shown in Table 6.4 which is just a variation on the matrix of Table 6.3.

TABLE 6.4 *The matrix of relative response strengths for two classes of words. Class A words all have sensitivities of α_A and biases of v. Class B words all have sensitivities of α_B and biases of 1. (After Ingleby. 1968)*

Stimulus alternative ↓		Response alternative →							
		Class A words				Class B words			
		1	2	3... N_A		1	2	3... N_B	
	1	$\alpha_A v$	v	v	v	1	1	1	1
	2	v	$\alpha_A v$	v	v	1	1	1	1
Class A words	3	v	v	$\alpha_A v$	v	1	1	1	1
	N_A	v	v	v	$\alpha_A v$	1	1	1	1
	1	v	v	v	v	α_B	1	1	1
	2	v	v	v	v	1	α_B	1	1
Class B words	3	v	v	v	v	1	1	α_B	1
	N_B	v	v	v	v	1	1	1	α_B

The situation represented in Table 6.4 can also be depicted in terms of the underlying distributions of signal and noise shown in Figures 6.4 and 6.5. Figure 6.4 shows the states of these distributions when a class B item has been presented as a stimulus. Incorrect class B items (noise) are normally distributed around a mean of 0. The correct item has a mean value of d'_B. The bias on class A items which are all incorrect, is represented by giving their distributions a mean value of k, somewhat above 0. The greater the bias to class A, the further above 0 will k lie. Figure 6.5 shows the distributions when a class A item has been used as the stimulus. Again class B items, which are all incorrect, have a mean value of 0 and the incorrect class A items have a mean value of k. The mean of the correct class A item is determined by two components. The first of these is the stimulus effect, d'_A, and the second, the response bias of size k. These add together so that the distribution mean will lie at a distance $d'_A + k$ above 0. Relating these figures to the choice theory matrix in Table 6.4 it can be seen that d'_B will be equal to $2 \log \alpha_B$, d'_A will be equal to $2 \log \alpha_A$, and k will be equal to $\log v$.

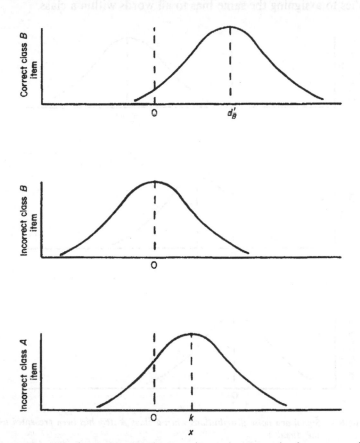

FIGURE 6.4   *Signal and noise distributions when a class B item has been presented as the stimulus*

Some questions arise about the assumptions underlying the construction of Table 6.4. Firstly, what right have we to assume that the αs for all items within a particular class are equal? The

143

answer is, none at all. It is quite likely that some class *A* items are more discriminable than others. However, the hypothesis being tested is that on the average α_A will be larger than α_B. We do not wish to obtain an α value for each member of a particular class but merely the average α for the whole class. The same argument also applies to assigning the same bias to all words within a class.

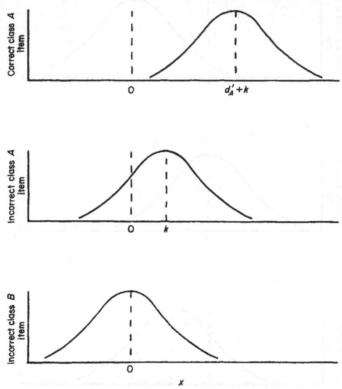

FIGURE 6.5 *Signal and noise distributions when a class A item has been presented as the stimulus*

A second question is, why is the bias put on class *A* words? Surely we cannot know before the experiment whether observers are biased in one direction or the other? And another supplementary question might be, why put no bias on the class *B* items? Would

144

it not be more appropriate to put a bias of v_A on class A items and one of v_B on class B items?

The general answer to these questions is that there is no way of obtaining an absolute measure of response bias in this situation. The measure v is a relative one. If the observer is biased towards class A responses then, to the same degree, he is biased away from class B responses. Table 6.4 is a matrix of *relative* response strengths. It shows how much more (or less) an observer is inclined to make a class A response rather than a class B response. So it does not matter whether the bias is put on class A or class B. If it turns out that observers are more willing to make class A responses than class B responses the value of v will be greater than 1. If observers show no more bias to one class than to the other, v will be equal to 1. If the bias is to class B, v will be less than 1. This also answers the question: why give a bias of 1 to class B and not v_B? As bias to class A is being measured relative to class B it does not matter what value of v is put on the class B words. A value of 1 is convenient for the purposes of calculation but any other number, greater than 0 would have done just as well.

The same argument applies to the pictorial representation of the matrix given in Figures 6.4 and 6.5. There the bias on class A items was represented by giving them a mean value of k in the absence of any stimulus effects. If observers were more biased in favour of class A responses than class B responses, k would have been positive. If there was no bias k would have been 0, and had there been a bias to class B, k would have been negative.

In an experiment, an observer may have been presented with an equal number of class A and class B items in random order and been asked to say, after each presentation, what the stimulus word was. Any response given must fall into one of three categories:

(a) It may be a correct response.
(b) It may be incorrect and from the same class as the stimulus.
(c) It may be incorrect and from a different class to the stimulus.

Each of these three possibilities has its response strength written in Table 6.4. Table 6.5 is a condensed version of Table 6.4 and it shows the relative response strengths for the three types of responses listed above, both for class A and class B stimuli.

145

TABLE 6.5 *A condensed version of Table 6.4 showing the relative strengths of the three possible types of response to stimuli from the two classes*

| | | Response alternative | | |
		Correct response	Any Class A error	Any Class B error
Stimulus alternative	Class A	(a) $\alpha_A v$	(b) v	(c) 1
	Class B	(d) α_B	(e) v	(f) 1

It should be possible to determine experimentally the proportions of responses made by an observer which fall into the three response categories and, by some simple multiplication, to find α_A, α_B, and v. One other thing has to be taken into account before this can be done. The large matrix of Table 6.4 shows that there are unequal numbers of class A and class B items; N_A of the former and N_B of the latter. Cell b of Table 6.5 shows that the response strength of any individual class A item which was not the correct response, is v. However, in the experiment no measure of individual response strengths has been obtained. We merely have the proportion of all class A errors to a class A stimulus. That is to say, the observed proportion of these class A errors is a measure of the response strength of all the $N_A - 1$ class A items which were not the correct item. Similarly, although cell c shows the response strength of any one class B response to a class A stimulus, the proportion of class B errors observed in the experiment represents the response strength of all the N_B class B items when a class A stimulus has been presented. The same applies to cells e and f. The proportion in e is the response strength of the

TABLE 6.6 *The table of relative response strengths modified to take into account the different numbers of Class A and B items*

| | | Response alternative | | |
		Correct responses	Class A errors	Class B errors
Stimulus alternative	Class A	(a) $\alpha_A v$	(b) $v(N_A - 1)$	(c) N_B
	Class B	(d) α_B	(e) $v N_A$	(f) $N_B - 1$

146

N_A incorrect class A items, and the proportion in f is the response strength of N_B-1 class B items. Table 6.6 is a modified version of Table 6.5, taking into account the different sizes of the two classes. It is this table which represents the proportions of different types of responses obtained experimentally.

Table 6.6 can now be used to estimate α_A, α_B, and v. Assuming that the values of N_A and N_B are known, and if a, b, c, d, e and f are the observed proportions of correct responses and errors for the cells of Table 6.6, then

$$\alpha_A = \frac{a(N_A-1)}{b}, \tag{6.11}$$

$$\alpha_B = \frac{d(N_B-1)}{f}. \tag{6.12}$$

It can be checked that these formulae are correct by substituting in them for a, b, d and f from Table 6.6.

There are two ways of determining v. First, the responses to class A stimuli can be used so that

$$v_a = \frac{bN_B}{c(N_A-1)}. \tag{6.13(a)}$$

Second, v can be found from responses to class B stimuli;

$$v_b = \frac{e(N_B-1)}{fN_A}. \tag{6.13(b)}$$

Notice that, although these two estimates of v have been called v_a and v_b, both are measures of the bias on class A items based on different data; v_b is not a measure of bias to class B. Therefore both formulae should give v values in the same direction; both greater than 1 if the bias is towards class A and both less than 1 if the bias is away from class A. It may not be the case however that the two formulae give v values of exactly the same size.

A SUMMARY OF THE PROCEDURE FOR AN OPEN-ENDED TASK

Having presented the rationale for the open-ended task, it may

147

be helpful to look at the sequence of steps an experimenter will need to go through if he is to use this method for estimating sensitivity and bias. For extra information about the practical details involved in running an experiment of this type, Broadbent (1967) and Broadbent and Gregory (1967) are useful sources of information. Here now is a skeleton design for an experiment. It will be assumed that the experimenter is interested in measuring sensitivity and bias for two classes of words, A and B, in a perceptual or memory experiment.

Definition of stimulus-response classes

The first step is to define precisely the criteria for including an item in each of the two classes. This is not only necessary for selection of stimuli; when the data have been collected, errors will have to be classified according to their response classes. If, as Broadbent was, we are interested in sensitivity and bias differences between common and uncommon words, class A, the common words might be defined as all words listed in the Oxford English Dictionary as common nouns and which occur in the AA class of the Thorndike-Lorge (1944) word count. Class B, uncommon words, might be defined as all common nouns in classes 10 to 49 of the Thorndike-Lorge count. This is an easy example for the experimenter to deal with as all he needs is a dictionary and the Thorndike-Lorge list to decide if a particular word is to be included in the relevant classes.

Suppose however, that the experimenter is concerned with differences between concrete and abstract nouns. Deciding whether a word is common or not is easy enough but how is its degree of concreteness to be determined? Assuming, for the moment, that there are no suitable published norms for this characteristic, the experimenter will need to do some preliminary work before beginning the experiment proper. One way of assessing concreteness would be to present a panel of judges with a list of randomly selected nouns (we may wish to place some restrictions on the frequency range from which these are to be selected and the Thorndike-Lorge count will be useful here) and ask them to rate on a five-point scale the degree of concreteness of each noun. A concreteness score for each item can be obtained and classes A and B can be defined as groups of nouns which receive scores within particular ranges. An example of this procedure, in this case applied to the determination of whether words were emotional or neutral, can be seen in Broadbent and Gregory (1967).

Verbal stimuli have been scaled along a number of dimensions, and the experimenter may save himself some time if he consults Runquist (1966) who lists some of the sources of normative data on a number of characteristics.

Size of stimulus-response classes

As N_A and N_B, the sizes of the classes whose sensitivities and biases are to be measured, are needed for the calculations, these must be found next. At this stage a problem may occur. Often it may not be possible to determine the absolute sizes of the two stimulus-response classes. Instead, the only information available may be the relative sizes of the two classes. For example, a sample of nouns may have been rated for concreteness-abstractness, and of that sample a proportion p met the criterion for being included in the concrete class while a proportion q met the criterion for inclusion in the abstract class. The ratio of p to q gives the ratio N_A/N_B, but from this alone the absolute sizes of N_A and N_B cannot be deduced. If this state of affairs occurs, some modifications need to be made to formulae (6.11), (6.12) and (6.13) before α and v values can be calculated. If the ratio N_A/N_B is called R_{AB}, the revised formulae are as follows:

$$\alpha_A^* = \frac{a}{b} R_{AB}, \tag{6.14}$$

$$\alpha_B^* = \frac{d}{f}, \tag{6.15}$$

$$v_a = \frac{b}{c} \cdot \frac{1}{R_{AB}}, \tag{6.16(a)}$$

$$v_b = \frac{e}{f} \cdot \frac{1}{R_{AB}}. \tag{6.16(b)}$$

where $R_{AB} = N_A/N_B$, and a, b, c, d, e and f are the cell entries of Table 6.6.

The following points should be noted about these formulae. First,

149

it is assumed that the sizes of the classes A and B are relatively large. To take (6.16(a)) as an example, if the values of b, c and R_{AB} are substituted, the expression $v_a = [v(N_A-1)/N_B][N_B/N_A]$ is obtained. The N_Bs cancel out to leave $v_a = v(N_A-1)/N_A$. If N_A is large, so that it is nearly equal to N_A-1, the terms on the right-hand side of the expression reduce to v. This is a feature of all the above formulae.

Second, the sensitivity measures α_A^* and α_B^* are actually equal to α_A/N_B and α_B/N_B as can be seen by substituting in (6.14) and (6.15). As both have been divided by the same constant N_B, they can be compared with one another, but the true α values can only be found if the absolute size of N_B is known.

Collection of data

The standard procedure has been to present to the observer an equal number of class A and class B items with the observer responding after each item has been presented. Naturally, in the presentation trials, items of type A and type B should occur in random order.

The number of trials needed to obtain α and v values from the raw data will depend both on the ease with which stimuli can be discriminated, and the magnitude of the response bias. To obtain α values for the two classes, some errors will have to be obtained in cells b and f of Table 6.6. To obtain values of v_a and v_b, errors must occur in cells b, c, e and f. If sensitivity is good, most responses may occur in cells a and d, and errors may fail to occur in the remaining cells. If there is a strong bias to one type of response, errors from the other class may be rare. The experimenter should reckon on having to present at least twenty-five to fifty stimuli from each class to ensure that some responses occur in each cell.

It sometimes happens that an observer fails to respond to some stimuli in an open-ended task. Although this decreases the data from which the detection measures are to be calculated, Ingleby (1968) has shown that these omissions do not distort the measures.

Treatment of data

Before α and v can be calculated, the observer's responses must be classified according to whether they were correct, or class A or B errors, so that a matrix like Table 6.6 can be drawn up. This means that the criteria used to define the two classes must be applied to each error. This may simply involve looking up a set of prepared norms or may require the errors to be rated by a panel of judges. Broadbent's (1967) study of the word-frequency effect is an example of the former procedure, and Broadbent's and Gregory's (1967) experiment on perception of emotional and neutral words, an example of the latter. Any response which does not meet the criteria for inclusion in the two classes is discarded, and does not enter into the subsequent analysis.

Depending on whether the absolute sizes of N_A and N_B are known, or whether the ratio N_A/N_B is the only information available, either formulae (6.11) to (6.13) or formulae (6.14) to (6.16) are then used to find α and v values. Before statistical analysis it may be desirable to transform the detection scores, as the occurrence of extreme values of α and v is not uncommon. The appropriate transformations to use are $\log \alpha$ and $\log v$. It will be remembered that these log values are approximately linearly related to d' and x of signal detection theory.

Generally each observer will have provided α values for both classes of stimuli so that significance tests, like the related-samples t-test or the Wilcoxin Matched-pairs test, will be appropriate for testing sensitivity differences. Although two measures of v may have been obtained from each observer, the question at issue is not whether v_a differs significantly from v_b, as both these scores are independent measures of the same response bias. The hypothesis normally under consideration is whether v differs significantly from 1 as a v of 1 is indicative of no bias to one response class or the other. The significance test to be used when there are just two response classes is a one-sample test. For example a one-sample t-test could be used to determine whether a set of $\log v$ scores differed significantly from 0. As an alternative, a sign test could be used to see if a significant number of v scores exceeds or falls below 1.

Interpretation of results

It must be remembered that the measures $2 \log \alpha$ and $\log v$ are only approximations to d' and x. It is not always the case that changes in the choice theory measures have a one to one correspondence to changes in the detection theory measures. By computer simulation

151

L

Ingleby (1968) has found that the following errors occur in the open-ended task:

(a) If there is a bias to one class of responses then sensitivity for that class will be underestimated. That is, the α value for the class will be less than the d' obtained from detection theory.

(b) If there is a bias away from a class of items then sensitivity for that class may be overestimated by using α as the sensitivity index.

(c) When there is no bias, α is an unbiased estimate of d', and when there is no difference in sensitivity, $\log v$ is an unbiased estimate of x.

The implication of this for interpreting results is if there is a significant bias to one class and it is also found that α for that class is significantly lower than for the other class, then the difference in sensitivity cannot be trusted. If the bias is away from the class or is zero, and it still transpires that the class's sensitivity is lower than that of the other, then the difference is a real one.

Extension to cases with more than two stimulus-response classes

All that has been said in the preceding sections can be applied to an experiment involving the use of stimuli from more than two classes. As an example the matrix of response strengths for three classes is shown in Table 6.7. It is like the two-class case of Table 6.6 but now there are three sensitivity components α_A, α_B, and α_C. The bias on class C stimuli has been fixed at 1 so that the biases on class A and B, v_A and v_B, are being measured relative to class C. Note that v_A and v_B in Table 6.7 are two different biases and not just independent measures of the same bias as v_a and v_b were.

TABLE 6.7   *The matrix of relative response strengths for an open-ended task involving three stimulus-response classes*

		Response alternative			
		Correct responses	Class A errors	Class B errors	Class C errors
Stimulus alternative	Class A	(a) $\alpha_A v_A$	(b) $v_A(N_A-1)$	(c) $v_B N_B$	(d) N_C
	Class B	(e) $\alpha_B v_B$	(f) $v_A N_A$	(g) $v_B(N_B-1)$	(h) N_C
	Class C	(i) α_C	(j) $v_A N_A$	(k) $v_B N_B$	(l) N_C-1

In the three-class case the sensitivity and bias measures are obtained by multiplication in the same way as for the two-class case. If the absolute sizes of the classes, N_A, N_B and N_C are known the formulae are as follows;

$$\alpha_A = \frac{a}{b}(N_A - 1) \tag{6.17}$$

$$\alpha_B = \frac{e}{g}(N_B - 1) \tag{6.18}$$

$$\alpha_C = \frac{i}{l}(N_C - 1) \tag{6.19}$$

$$v_{Aa} = \frac{b}{d} \cdot \frac{N_C}{N_A - 1} \tag{6.20(a)}$$

$$v_{Ab} = \frac{f}{h} \cdot \frac{N_C}{N_A} \tag{6.20(b)}$$

$$v_{Ac} = \frac{j}{l} \cdot \frac{N_C - 1}{N_A} \tag{6.20(c)}$$

$$v_{Ba} = \frac{c}{d} \cdot \frac{N_C}{N_B} \tag{6.21(a)}$$

$$v_{Bb} = \frac{g}{h} \cdot \frac{N_C}{N_B - 1} \tag{6.21(b)}$$

$$v_{Bc} = \frac{k}{l} \cdot \frac{N_C - 1}{N_B} \tag{6.21(c)}$$

Where $a, b, c, \ldots, l$ are the cells of Table 6.7.

Notice also that v_{Aa}, v_{Ab}, and v_{Ac} are all alternate ways of calculating the bias to class A items and that v_{Ba}, v_{Bb}, and v_{Bc} are alternate ways of calculating the bias to class B items.

If the absolute numbers of items in the classes are not known, sensitivity and bias measures can still be obtained if information is available about the relative proportions of items in the various classes. From these, the two ratios $R_{AC} = N_A/N_C$ and $R_{BC} = N_B/N_C$

can be found. The formulae for the α and v values then become:

$$\alpha_A^* = \frac{a}{b} R_{AC} \tag{6.22}$$

$$\alpha_B^* = \frac{e}{g} R_{BC} \tag{6.23}$$

$$\alpha_C^* = \frac{i}{l} \tag{6.24}$$

$$v_A = \frac{f}{h} \cdot \frac{1}{R_{AC}} \tag{6.25}$$

$$v_B = \frac{c}{d} \cdot \frac{1}{R_{BC}} \tag{6.26}$$

Using these formulae means that $\alpha_A^* = \alpha_A/N_C$, $\alpha_B^* = \alpha_B/N_C$, and $\alpha_C^* = \alpha_C/N_C$. The formulae for v_A and v_B have been chosen so that the ratio exactly cancels out all terms except the v one.

This example should make obvious the way in which choice theory analysis can be extended to open-ended tasks with any number of stimulus-response classes. The basic procedure involves the writing of a matrix of relative response strengths like that of Table 6.7 and then devising expressions like the above formulae to isolate each α and v term in the matrix.

Problems

1. Use the logistic approximation to the normal distribution to find the following:

   (a) The proportion of the area above a criterion 0·50 S.D.s below the distribution mean.
   (b) The proportion of the area below a criterion 0·75 S.D.s below the distribution mean.
   (c) The proportion of the area above a criterion 0·60 S.D.s above the distribution mean.
   (d) The proportion of the area below a criterion 0·40 S.D.s above the distribution mean.

2. Using the choice theory approximations to signal detection

theory find :
(a) $P(S|s)$ and $P(S|n)$ when $\alpha = 6$ and $v = 4$.
(b) $P(N|s)$ and $P(N|n)$ when $\alpha = 2$ and $v = 0.5$.
(c) α and v when $P(S|s) = 0.70$ and $P(S|n) = 0.20$.
(d) $P(S|s)$ and $P(S|n)$ when the distance between signal and noise means is 1 S.D. and the criterion is 0.5 S.D.s below the noise distribution mean.
(e) The distance between signal and noise means, and the distance of the criterion from the noise distribution mean when $P(S|s) = 0.50$ and $P(S|n) = 0.30$.
(f) $\log v$ when $2 \log \alpha = 2$ and $P(S|s) = 0.25$.

3. A population of 250 items is divided into three classes. The number of items in each class is as follows :

Class :	A	B	C
Number of items :	175	50	25

From each of classes A and B, 50 items are selected at random and presented to an observer in a perceptual task in which he makes an open-ended response after the presentation of each item. His results are given below. From them find :
(a) α_A.   (b) α_B.
(c) v_a and v_b assuming that the bias on class B items is fixed at $v = 1$.

		Correct responses	Class A errors	Class B errors	Class C errors
Stimulus	Class A	15	33	1	1
item	Class B	5	43	2	0

4. The following table shows the percentage of responses which were either correct or errors in an open-ended task in which stimuli from four classes were presented an equal number of times. From the data find :
(a) α_A^*, α_B^*, α_C^*, and α_D^*.
(b) v_A, v_B, and v_C.
The ratios of the number of items in classes A, B and C to the number of items in class D are known to be $N_A : N_D = 2:1$, $N_B : N_D = 4:1$; $N_C : N_D = 0.5:1$.

155

In your calculations fix the bias on class B items at $v = 1$ and use N_D as the denominator for the α^* values.

		Observer's response				
	Class	Correct responses	Class A errors	Class B errors	Class C errors	Class D errors
	A	64%	9%	2%	20%	5%
Stimulus	B	64%	10%	1%	20%	5%
item	C	64%	10%	2%	19%	5%
	D	64%	10%	2%	19%	5%

Chapter 7

THRESHOLD THEORY

Now that the methods of signal detection theory have been presented, it remains to put them into the context of experimental psychology. To do this it will be necessary to discuss a number of procedures called *psychophysical methods* which have been in use in psychology since the time of Fechner in the second half of the nineteenth century. Psychophysical methods are a number of more or less standard techniques for collecting data in perceptual and learning tasks. Yes–no, rating scale, and forced-choice tasks are psychophysical methods, but the measures which have been extracted from them by signal detection theory differ from those of so-called classical psychophysics.

CLASSICAL PSYCHOPHYSICS
(Dember, 1964; Green & Swets, 1966, 117–48; Guilford, 1936; Torgerson, 1958; Swets, 1961)

The basic assumption of classical psychophysics is the existence of a *threshold*, and the basic aim of the classical psychophysical methods is to measure this threshold. The threshold, or *limen*, can be thought of as a barrier which a stimulus must exceed before it can be perceived. To fall back on the terminology we have been using in previous chapters, unless the value of x, the evidence variable, exceeds some value L, of the threshold, then the observer will not be able to detect the presence of the stimulus. The stimulus will be subliminal. Thus all values of x below L will be indistinguishable from each other, and will have no sensory effects.

Another feature of this type of threshold theory is that it assumes

157

the threshold to be sufficiently high for it to be never or rarely exceeded by the effects of noise alone. Hence the threshold must be set well up beyond the mean of the noise distribution if false alarms are to be rare or non-existent events. Green & Swets (1966) note that if Gaussian distributions of signal and noise are assumed, the threshold will need to be set at about three S.D.s above the mean of the noise distribution. Consequently, classical threshold theory is often called *high threshold theory*.

From the previous chapters it will be realized that the assumptions of detection theory are quite different. It has been taken for granted that an observer can set his criterion at any value of *x*, responding *S* to all values of the evidence variable which exceed this value, and *N*

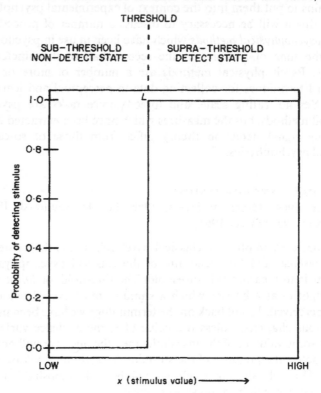

FIGURE 7.1 *Probability of detecting a signal whose value falls below and above the threshold. L—the ideal case*

158

to all values of x which fall below it. This assumes that the observer can discriminate between different values of x on all regions along the x-axis on which signal and noise distributions are scaled. Also, according to detection theory, the value of x which the observer chooses as his criterion will depend on such considerations as the relative probabilities of occurrence of signal and noise, and rewards and costs of different stimulus-response outcomes. There may be occasions when the criterion is set three S.D.s above the noise distribution mean, as may occur in high threshold theory, but there may equally well be times when the criterion is set three S.D.s below the mean of the noise distribution. Detection theory assumes that there is no signal which is too small to result in some discriminable sensory effect. that there is no difference between two signals too small to be detected, and that the criterion can be set at any distance from the noise and signal distribution means, allowing false alarms to occur with probabilities approaching either zero or unity.

As detection theory and high threshold theory stand in contrast to one another, it is important to know which is the more valid description of perceptual decision processes. A number of means have been used to evaluate the two theories and these will be discussed after the methods of classical psychophysics have been described.

The threshold

Ideally, the value $x = L$ of the high threshold should be a constant. The probability of detecting a signal with a value of x_i when $x_i \geqslant L$ or $x_i < L$ will be as in Figure 7.1.

Below $x_i = L$, the probability of detection is 0; above $x_i = L$, the probability of detection is 1.

Empirical attempts at determining the value of L do not yield a single precise value of the evidence variable, but a distribution of threshold values. The threshold can be thus considered as varying randomly about a mean value according to a Gaussian distribution. This mean value will be called L. The probability of detecting a signal with a value of x_i when $x_i \geqslant L$ or $x_i < L$ is illustrated in Figure 7.2. In this case there is no sharp transition from a non-detect to a detect state, but the probability of making a detection as a function of x assumes an s-shaped curve which is called a *normal*

159

ogive. It can be seen that when $x_i = \bar{L}$, the probability of detecting a signal equals 0·5. Operationally, therefore, the threshold is defined as the value of *x* at which the signal is detected 50 % of the time.

Two main types of threshold have been of interest to classical psychophysics. The first of these, the *absolute threshold*, is the least amount of energy capable of being detected by the perceptual

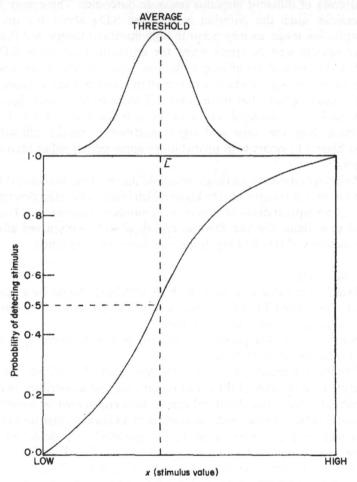

FIGURE 7.2   *Probability of detecting a signal whose value falls below and above the average value.* $\bar{L}$, *of a threshold which varies—the empirical case.*

160

system. The second of these, the *difference threshold*, is the smallest change of a stimulus which can be detected, assuming that the stimulus is itself above the absolute threshold. The absolute threshold is thus supposed to represent the absolute sensitivity of the perceptual system, and the difference threshold, its powers of resolution. Operationally, the absolute threshold is that value of the stimulus variable which is detected on 50% of occasions, and the difference threshold is that difference between a pair of stimuli which can be perceived on 50% of occasions.

Threshold measurement

Dember (1964) has classified the classical psychophysical methods according to the means by which stimuli are selected for presentation, and the type of response required from the observer. Two possibilities are available for stimulus selection:

(a) *The method of limits.* The stimulus selected for the current trial depends on the response made by the subject on the previous trial. In this case the experimenter may select, at first, a stimulus whose value is well above the presumed threshold. This is presented to the observer who will probably be able to detect it. The stimulus value is then gradually decreased, and the subject attempts to detect it until a point is reached where the stimulus is too weak to be detected. This is a *descending* series of trials. Then, starting with a stimulus whose value falls below the threshold, the experimenter conducts a series of *ascending* trials which continue until the stimulus reaches a value where the observer can detect it.

(b) *The method of constant stimuli.* The stimulus selected for the current trial is independent of responses made by the subject on previous trials. In this method the experimenter will characteristically choose a number of stimuli whose values are in the region of the presumed threshold. These may then be presented a number of times each, either in random order, or in blocks of the same stimulus. The observer indicates whether or not each stimulus is detectable, so that the experimenter can obtain, for each stimulus value, the number of times it was detected.

Also, there are two possible ways of obtaining responses from the

161

observer, a yes–no task or an *m*AFC task. As in the detection theory methods, the yes–no task in classical psychophysics requires the observer to indicate whether or not he detected the stimulus on that trial. Blank trials (that is those in which a stimulus is not presented) may be interspersed amongst the stimuli, not, as in detection theory, to obtain a measure of the false alarm rate, because high threshold theory does not admit the possibility of false alarms, but to check whether the observer is spuriously inflating the proportion of detections by chance guessing.

In the forced-choice task the observer in an absolute threshold experiment will be presented with two or more intervals, one of which contains the stimulus, and the others of which are blank. As in detection theory experiments, he must select the stimulus interval. In a difference threshold experiment, the observer may be presented with a pair of intervals, both containing different stimuli, and asked to indicate which interval contained the greater stimulus.

With two methods of stimulus selection and two techniques for collecting responses there should be four possible types of psychophysical experiment, but, as Dember points out, the method of limits cannot be easily combined with forced-choice procedure. The other three combinations are possible ones.

HIGH THRESHOLD THEORY AND THE YES–NO TASK

A comparison of the respective merits of high threshold theory and detection theory can now be begun by looking first at the yes–no task (this, of course, also includes the rating scale task). There are a number of points of difference (see Green & Swets, 1966, Ch. 5; Swets, 1961) but we will focus on differences between the ROC curves predicted by the theories and on a function called the *a posteriori* function, which will be described presently.

The ROC *curve for high threshold theory*
Figure 7.3 is a picture of the assumptions of high threshold theory. In the example there, it is assumed that the observer is presented with 100 signal and 100 noise trials. Of the 100 signals, 60 lie above L, the threshold, giving a true hit rate, $P_T(S|s) = 0.60$. However, for all events which fall below the threshold, and this includes all

162

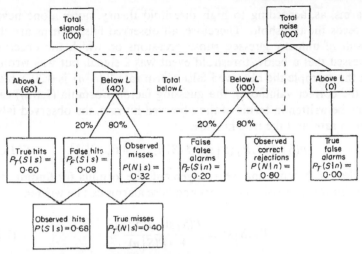

FIGURE 7.3   *An illustration of the assumptions of high threshold theory showing the way in which the true hit rate is inflated by the effects of chance guessing. In this example it is assumed that 20% of events falling below L, the threshold, are guessed as signals*

100 noise events in addition to the 40 signals, the observer guesses that 20% of these are signals. Thus by chance guessing he increases his total hits by $0.20 \times 40 = 8$. The observed hit rate, $P(S|s)$, is thus the sum of true hits plus false hits $P_F(S|s)$. Hence

$$P(S|s) = P_T(S|s) + P_F(S|s) \qquad (7.1(a))$$

If we knew what the false hit rate was, it would be possible to deduce true hits from observed hits. Notice first that the false hit rate is the proportion of true misses (i.e. signals which fall below L) which are correctly guessed as being signals. Formula (7.1(a)) can therefore be rewritten as follows:

$$P(S|s) = P_T(S|s) + g[1 - P_T(S|s)] \qquad (7.1(b))$$

In this formula $[1 - P_T(S|s)]$ is the true miss rate, in this example, 0.40; and g is the guessing factor, in this example 20%. Now all we need to know is the value of g to determine the true hit rate from the observed hit rate. The value of g is, of course, the observed false alarm rate. In Figure 7.3 it can be seen that there are no true false alarms, as, according to high threshold theory, noise alone never exceeds the threshold. Therefore, all observed false alarms are the result of unlucky guesses, those occasions on which the observer guessed that a below-threshold event was a signal, but was wrong. In the example the observed false alarm rate, $P(S|n)$, is 0.20, which is the correct value of g, the guessing factor. Formula (7.1(b)) can now be written in terms of the observed hit rate, the observed false alarm rate, and the true hit rate.

$$P(S|s) = P_T(S|s) + P(S|n) \cdot [1 - P_T(S|s)]. \qquad (7.1(c))$$

Now with some final rearrangement, the formula for finding the true hit rate from observed hits and false alarms can be written:

$$P_T(S|s) = \frac{P(S|s) - P(S|n)}{1 - P(S|n)}. \qquad (7.2)$$

This is the well-known *correction for chance guessing* for a yes–no task. It expresses the true hit rate, $P_T(S|s)$, as a function of observed hits and observed false alarms. As d' is the index of sensitivity in detection theory it can be seen that $P_T(S|s)$ is the sensitivity measure in high threshold theory. Its value will be equal to 0.5 for a stimulus at the threshold, greater than 0.5 for above-threshold stimuli, and less than 0.5 for below-threshold stimuli. Changes in response bias may produce changes in $P(S|s)$ and $P(S|n)$, the observed hit and false alarm rates, but $P_T(S|s)$ should remain constant as bias varies.

The ROC curve for high threshold theory can be constructed with the aid of formula (7.1). Given a true hit rate of $P_T(S|s) = 0.5$, the formula can be used to generate a set of values of $P(S|s)$ and $P(S|n)$ which represent different degrees of response bias, but the same degree of sensitivity. This set of values will define the path of the ROC curve. Assume, for a start, that the observer does not guess any sub-threshold events as signals. The guessing factor g, which is equal to $P(S|n)$, will thus be 0.0. Substituting 0.0 for $P(S|n)$, and 0.5 for $P_T(S|s)$ in (7.1), the observed hit rate $P(S|s)$ is seen to be 0.5. This gives the first point for the ROC curve. Adopting a different degree of bias is tantamount to changing the value of g. Assume this is increased to 0.2. Thus $P(S|n)$ will be equal to 0.2, $P_T(S|s)$ will remain unchanged at 0.5, and substituting in (7.1), $P(S|s)$ will

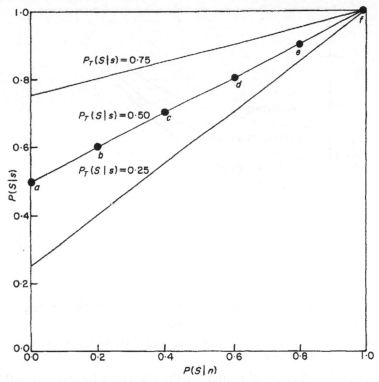

FIGURE 7.4 ROC *curves for high threshold theory, for the yes–no task, plotted on P-scales*

TABLE 7.1 *Calculation of points for the Yes–No* ROC *curve for a true hit rate of* 0·5

Criterion	True hits $P_T(S/s)$	Guessing factor $g = P(S\|n)$	Observed hits $P(S\|s) = P_T(S\|s) + P(S\|n)[1 - P_T(S\|s)]$
a	0·5	0·0	0·5
b	0·5	0·2	0·6
c	0·5	0·4	0·7
d	0·5	0·6	0·8
e	0·5	0·8	0·9
f	0·5	1·0	1·0

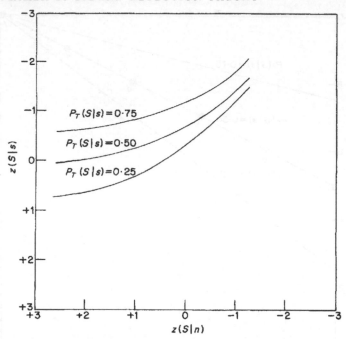

FIGURE 7.5 *A double-probability plot of the* ROC *curves for high threshold theory shown in Figure 7.4*

become 0.6. A series of points for the ROC curve for $P_T(S|s) = 0.5$ is shown in Table 7.1, and the curve is plotted in Figure 7.4. It is a straight line running from $P(S|n) = 0.0$ and $P(S|s) = P_T(S|s)$ to $P(S|n) = P(S|s) = 1.0$. A number of other ROC curves for various values of $P_T(S)|s)$ are shown in Figure 7.4, and a double-probability plot of these curves is shown in Figure 7.5.

As high threshold theory and signal detection theory predict different ROC curves, it should be easy to choose between them on the basis of experimental data. There are two types of information that empirical ROC curves might yield which would help decide between the two theories. First, and obviously, the shapes are different. Second, the best theory will be the one whose index of sensitivity remains constant when the stimulus conditions are held constant but response conditions varied. That is to say, if the levels of signal and noise are not changed, but the observer is tested under

166

different degrees of response bias, detection theory will be preferable to high threshold theory if d' remains constant, but $P_T(S \mid s)$ varies, and high threshold theory will be preferred if the converse holds.

In neither case does high threshold theory seem to be an adequate description of the decision process. Swets (1961), and Swets, Tanner & Birdsall (1961) have found that ROC curves from both yes–no tasks and rating scale tasks have the shape predicted by detection theory and not that predicted by high threshold theory. Swets, Tanner & Birdsall used a visual detection task in which a brief circular spot of light (the signal) was presented on a uniformly illuminated background. Observers participated in a series of sessions in which four levels of signal intensity were used. Yes–no procedure was used, and different points for the ROC curves were obtained by varying the probability of signal and the payoffs for different stimulus-response combinations. The shapes of the ROC curves were of the type predicted by signal detection theory, but not of the type predicted by high threshold theory. Also when a rating scale task was used with the same type of stimulus situation, the results again favoured signal detection theory. In this rating scale task, observers were asked to classify each stimulus event according to six categories of how probable it was that a stimulus was a signal. Category 1 corresponded to a subjective probability of 0·00 to 0·04 that the stimulus was a signal, category 2 to a subjective probability of 0·05 to 0·19 and categories 3 to 6, to subjective probabilities of 0·20 to 0·39, 0·40 to 0·59, 0·60 to 0·79 and 0·80 to 1·00, respectively. ROC curves of the shape predicted by signal detection theory have been reported in many other areas of psychology, for instance, in the study of human vigilance (Broadbent & Gregory, 1963a); selective attention (Broadbent & Gregory, 1963b); speech perception (Pollack & Decker, 1958); perceptual defence (Dandeliker & Dorfman, 1969); short-term memory for paired-associates (Murdock, 1965); and memory for serial positions (Wickelgren & Norman, 1966).

In addition it was said that detection theory would be preferred to high threshold theory if d' remained constant as response bias was changed and levels of signals and noise were held constant. Swets (1961) reports experiments on vision by Swets, Tanner & Birdsall, and on audition by Tanner, Swets & Green, where it was found that d' remained invariant with changes in response bias.

167

M

A posteriori probabilities for high threshold and signal detection theory

In describing the rating scale task in Chapter 2, it was said that an observer may be able to give more information about a stimulus than a simple yes–no decision as to whether it was signal or noise. The use of rating scale techniques assumes that observers are able to order events according to the likelihoods that they are signal. Values of x, the evidence variable, which yield large values of $l(x)$ should be rated confidently as being signals. Those values of x giving small values of $l(x)$ should be rated confidently as being noise. A graph could therefore be constructed showing the probability of detecting a signal for each category on the rating scale. If the data from Table 5.1 are used as an example, the probability of detecting a signal in category 1 will be the number of signals detected (159) divided by the total number of responses made in category 1 (159 + 2). For the five rating categories the graph of $P(s|R_j)$, the probability of there being a signal given that the observer has made response j on the rating scale, is shown in Figure 7.6. It can be seen that $P(s|R_j)$ increases monotonically as a function of confidence that the event was a signal, showing that the observer is able to order sensory events according to the likelihood that they are signals.

$P(s|R_j)$ is called an *a posteriori* probability, and the curve in Figure 7.6 is called an *a posteriori* function. The shape of this function differs for different theories about the decision process. If, as in detection theory, we assume that there is no signal too weak to be detected, and no difference between signals too small to be observed, the *a posteriori* function will be, as in Figure 7.6, a monotonically increasing function of confidence that the stimulus was a signal. The shape of the function for high threshold theory is quite different but before discussing this matter we will deal with the topic of *a posteriori* probabilities more thoroughly.

An observer in a detection task may be told at the beginning of the experiment that the probability of occurrence of signals, $P(s)$ will be 0·5. If, at any stage during the experiment, he is asked to say what he thinks will be the chances of a signal being presented on the next trial, he should say 50%. $P(s)$ is called the *a priori* probability of signal occurring.

Now assume that the subject knows the *a priori* probability of signal, and is then presented with a stimulus event which might be

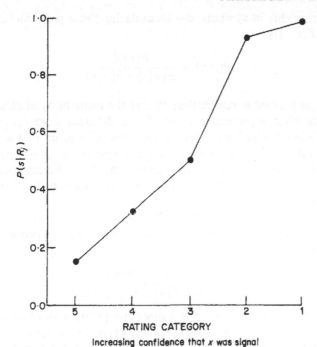

FIGURE 7.6 A posteriori *function for signal detection theory based on the data of Table 5.1*

either signal or noise. He is then asked to say, after the event has been presented, what is the probability that that event was a signal. If he has some information about the distribution of x, the evidence variable, he may be able to modify the *a priori* probability in the light of the extra evidence which has been provided.

We return to the example in Figure 1.1 and Table 1.1, and consider the case where an observer is given a piece of evidence where $x = 66$ in. and knows that $P(s) = P(n)$. The *a priori* probability of signal is simply $P(s)$ or 0.5. The probability of obtaining $x = 66$ in. from noise alone can be seen to be 4/16, and from signal + noise, 3/16. Therefore, the probability of a signal having been presented when $x = 66$ in. will be

$$\frac{3/16}{3/16 + 4/16} = 3/7.$$

169

Putting this in symbols, the formula for the *a posteriori* probability, $P(s|x)$ is

$$P(s|x) = \frac{P(x|s)}{P(x|s) + P(x|n)}. \tag{7.3}$$

It is important to realize that $P(s|x)$, the probability of obtaining a signal given a particular value of x, is different from $P(x|s)$, the probability of obtaining a particular value of x given that a signal has occurred. If this seems obscure, consider the probability of something being two years old, given that it is a human being as opposed to the probability of it being human given that it is two years old.

Formula (7.3) applies only to the special case where $P(s) = P(n)$. To take account of $P(s)$ and $P(n)$ having different *a priori* probabilities we must write

$$P(s|x) = \frac{P(s) \cdot P(x|s)}{P(s) \cdot P(x|s) + P(n) \cdot P(x|n)}. \tag{7.4}$$

The denominator of (7.4) is simply the probability of obtaining a particular value of x, or $P(x)$. In the example we have just been using, where x was 66 in. and $P(s)$ was equal to $P(n)$, the probability, $P(x)$, of obtaining $x = 66$ in., regardless of whether it was signal or noise is $3/16 + 4/16 = 7.16$. So, writing $P(x)$ in place of the denominator (7.4) becomes

$$P(s|x) = \frac{P(s) \cdot P(x|s)}{P(x)}. \tag{7.5}$$

This expression for the *a posteriori* probability is called Baye's Rule, which has been applied in a number of studies of the ways in which subjects revise subjective probabilities in decision-making tasks (Coombs, Dawes & Tversky, 1970).

Another re-arrangement of formula (7.4) shows how *a posteriori* probability is related to $l(x)$, the likelihood ratio. If (7.4) is inverted, it becomes

$$P(s|x) = \left(\frac{P(s)P(x|s) + P(n)P(x|n)}{P(s)P(x|s)} \right)^{-1}.$$

By separating out the terms in the numerator of this expression we obtain

$$P(s|x) = \left(1 + \frac{P(n)}{P(s)} \cdot \frac{P(x|n)}{P(x|s)}\right)^{-1}.$$

Of course $P(x|n)/P(x|s)$ which appears in the above expression is $1/l(x)$, the reciprocal of the likelihood ratio for x being signal. So we can now write

$$P(s|x) = \left(1 + \frac{P(n)}{P(s)} \cdot \frac{1}{l(x)}\right)^{-1}.$$

When this last expression is tidied up we get the formula for the *a posteriori* probability expressed in terms of the *a priori* probabilities of s and n, and the likelihood ratio.

$$P(s|x) = \frac{P(s)l(x)}{P(s)l(x) + P(n)}. \tag{7.6}$$

In rating scale tasks we do not deal directly with values of x, but with confidence categories. So instead of using $P(x|s)/P(x|n)$ as the expression for the likelihood ratio for a particular value of x, we will wish to use $P(R_j|s)/P(R_j|n)$ as the expression for the likelihood ratio for a particular confidence category. By substituting $P(R_j|s)/P(R_j|n)$ for $l(x)$ in (7.6) we obtain (7.7) which is the expression for $P(s|R_j)$, the *a posteriori* probability associated with a particular confidence category:

$$P(s|R_j) = \frac{P(s)P(R_j|s)}{P(s)P(R_j|s) + P(n)P(R_j|n)}. \tag{7.7}$$

The form of (7.7) closely resembles that of (7.4) and they can be treated as equivalent so long as it can be assumed that confidence is monotonic with x. A slight problem occurs here, as we know from Chapter 4 that x is not monotonic with $l(x)$ when variances of signal and noise are unequal, and there is no firm evidence to date as to whether observers base their decisions on the value of the likelihood ratio or simply on the value of the evidence variable itself. If likelihood ratios are the basis of a confidence judgement, then in theory, (7.7) will not always be an acceptable substitute for (7.4), but in practice, as has been pointed out before, the monotonic relationship

171

between x and $l(x)$ breaks down in the extreme tails of the distributions and these are not likely to be able to be explored in a real experiment.

The shape of the *a posteriori* functions for signal detection theory and for high threshold theory can now be considered. Figure 7.6 has already shown that *a posteriori* probability is a monotonic increasing function of confidence for signal detection theory.

In high threshold theory, any stimulus above the threshold will be seen reliably by an observer to be a signal, and should thus be allocated to the highest category of confidence. On the other hand, any event below the threshold, be it signal or noise, will, by definition, be indistinguishable from any other event. There should thus be no correlation between the confidence category to which a sub-threshold event is assigned and whether it was signal or noise. Any method of assigning sub-threshold events to confidence categories is, to all practical purposes, a random one, so that the *a posteriori* function for high threshold theory should be of the form shown in Figure 7.7. In this figure, category 1 is reserved for above-threshold events, and as these are all signals, $P(s|R_1) = 1$. The remaining confidence categories are used for sub-threshold events and signal and noise events are assigned to them at random so that $P(s|R_2) = P(s|R_3) = \ldots = P(s|R_j) = \ldots = P(s|R_N)$. The *a posteriori* function predicted by high threshold theory has not been reported in the literature, although the function predicted by detection theory has been reported by Swets, Tanner & Birdsall (1961) in a study of visual detection, and Broadbent & Gregory (1963a) in a study of vigilance. Norman & Wickelgren (1965) have reported an *a posteriori* function from a short-term memory experiment which is not of the form predicted by signal detection theory, but it is even less like the type of function predicted by high threshold theory. On these grounds, high threshold theory must again be rejected as an adequate description of decision processes in human perception and memory, while signal detection theory again holds up rather well in the light of experimental evidence.

FORCED-CHOICE TASKS

High threshold theory and signal detection theory also differ in

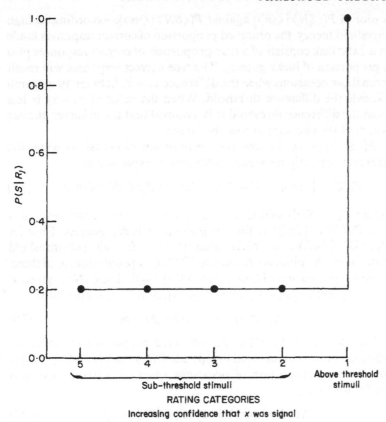

FIGURE 7.7   *The* a posteriori *function predicted by high threshold theory*

their predictions about forced-choice tasks. As in previous discussions of forced-choice tasks we begin with the 2AFC task and then generalize to the *m*AFC task.

The 2AFC ROC *curve for high threshold theory*

Egan, in an unpublished report, summarized by Green & Swets (1966, 337–41) has derived the ROC curve for the 2AFC task for high threshold theory. In Chapter 2 we saw that an observer in a 2AFC task has to choose between two hypotheses ⟨*sn*⟩ and ⟨*ns*⟩. He has two responses ⟨*SN*⟩ and ⟨*NS*⟩, and the 2AFC ROC curve consists of

173

a plot of $P(\langle SN \rangle | \langle sn \rangle)$ against $P(\langle SN \rangle | \langle ns \rangle)$. According to high threshold theory, the observed proportion of correct responses made in a 2AFC task consists of a true proportion of correct responses plus a proportion of lucky guesses. The true correct responses will result from those occasions when the difference $x_1 - x_2$ between two stimuli exceeds the difference threshold. When the value of $x_1 - x_2$ is less than the difference threshold it is assumed that the observer guesses which of the two stimuli was the larger.

$P(\langle SN \rangle | \langle sn \rangle)$, the observed proportion of occasions when the observer correctly responds $\langle SN \rangle$ can be expressed as:

$$P(\langle SN \rangle | \langle sn \rangle) = P_T(\langle SN \rangle | \langle sn \rangle) + P_F(\langle SN \rangle | \langle sn \rangle), \qquad (7.8)$$

where $P_T(\langle SN \rangle | \langle sn \rangle)$ is the true proportion of correct responses and $P_F(\langle SN \rangle | \langle sn \rangle)$ is the proportion of lucky guesses. First let $P_T(\langle SN \rangle | \langle sn \rangle) = p$. Next assume that for all sub-threshold differences, the observer responds $\langle SN \rangle$ to a proportion, g, of them. Then the proportion of times, $P_F(\langle SN \rangle | \langle sn \rangle)$, that $\langle SN \rangle$ is guessed correctly will be $g[1 - P_T(\langle SN \rangle | \langle sn \rangle)]$, or $g(1 - p)$. Therefore

$$P(\langle SN \rangle | \langle sn \rangle) = p + g(1 - p). \qquad (7.9)$$

Now if $P(\langle SN \rangle | \langle ns \rangle)$ is the observed proportion of occasions when the observer incorrectly responds $\langle SN \rangle$, and $P(\langle NS \rangle | \langle ns \rangle)$ is the observed proportion of occasions when he correctly responds $\langle NS \rangle$, we have

$$P(\langle SN \rangle | \langle ns \rangle) = 1 - P(\langle NS \rangle | \langle ns \rangle). \qquad (7.10)$$

$P(\langle NS \rangle | \langle ns \rangle)$ like $P(\langle SN \rangle | \langle sn \rangle)$ will consist of a true proportion of correct responses, $P_T(\langle NS \rangle | \langle ns \rangle)$, and a proportion of lucky guesses, $P_F(\langle NS \rangle | \langle ns \rangle)$, so that

$$P(\langle NS \rangle | \langle ns \rangle) = P_T(\langle NS \rangle | \langle ns \rangle) + P_F(\langle NS \rangle | ns \rangle). \qquad (7.11)$$

As the true proportion of $\langle NS \rangle$ responses should be equal to the true proportion of $\langle SN \rangle$ responses, $P_T(\langle NS \rangle | \langle ns \rangle) = P_T(\langle SN \rangle) \langle sn \rangle) = p$.

As g was the proportion of sub-threshold differences which was responded to as $\langle SN \rangle$, the proportion guessed as $\langle NS \rangle$ will be $1 - g$. Then the proportion of times, $P_F(\langle NS \rangle | \langle ns \rangle)$, that $\langle NS \rangle$

is guessed correctly will be $(1-g)\left[1-P_T(\langle NS\rangle|\langle ns\rangle)\right]$ so that

$$P_F(\langle NS\rangle|\langle ns\rangle) = (1-g)(1-p). \qquad (7.12)$$

If in (7.11), we substitute p for $P_T(\langle NS\rangle|\langle ns\rangle)$, and $(1-g)(1-p)$ for $P_F(\langle NS\rangle|\langle ns\rangle)$, we have

$$P(\langle NS\rangle|\langle ns\rangle) = p+(1-g)(1-p). \qquad (7.13)$$

Substituting the expression for $P(\langle NS\rangle|\langle ns\rangle)$ from (7.13) in (7.10) we obtain the following expression for $P(\langle SN\rangle|\langle ns\rangle)$

$$P(\langle SN\rangle|\langle ns\rangle) = g(1-p). \qquad (7.14)$$

Formulae (7.9) and (7.14) express the observed conditional probabilities $P(\langle SN\rangle|\langle sn\rangle)$ and $P(\langle SN\rangle|\langle ns\rangle)$ in terms of a true proportion of correct responses, p, and a guessing factor, g. In this respect they resemble (7.1) from which the yes–no ROC curve was constructed for high threshold theory. Analogously, the ROC curve for the 2AFC task predicted by high threshold theory can be constructed from formulae (7.9) and (7.14). The term p is the sensitivity measure for high threshold theory, and the value of g represents the degree of response bias. If we select a particular level of sensitivity, say $p = 0.4$, and several levels of response bias, say $g = 0.00, 0.33, 0.67, 1.00$, formulae (7.9) and (7.14) may be used to determine several points for a 2AFC ROC curve. These calculations are shown in Table 7.2 and the ROC curve is plotted in Figure 7.8. It can be seen that the 2AFC ROC curve for high threshold theory plotted on scales of P-units is a straight line with a slope of 1. This is in contrast to the predicted 2AFC ROC curve for signal detection theory, which is a straight line

TABLE 7.2 *Calculation of points for the 2AFC ROC curve of high threshold theory when the true proportion of correct responses equals 0.40*

Criterion	True proportion of correct responses p	Guessing factor g	$P(\langle SN\rangle|\langle sn\rangle)$ $=p+g(1-p)$	$P(\langle SN\rangle|\langle ns\rangle)$ $=g(1-p)$
a	0.40	0.00	0.4	0.0
b	0.40	0.33	0.6	0.2
c	0.40	0.67	0.8	0.4
d	0.40	1.00	1.0	0.6

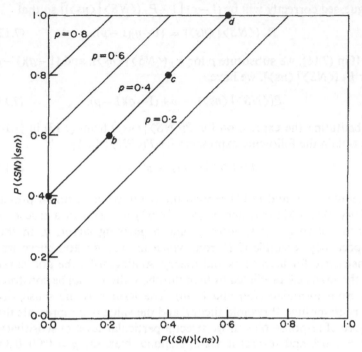

FIGURE 7.8   2AFC ROC curves for high threshold theory

with a slope of 1 when plotted on a double-probability scale, but a symmetrical bow-shaped curve when plotted on scales of P-units.

2AFC operating characteristics of the type predicted by high threshold theory are yet to be reported. ROC curves of the type predicted by signal detection theory have been reported by Schulman & Mitchell (1966) and Markowitz & Swets (1967) in studies of auditory detection. In an unpublished study carried out at the Department of Psychology, Adelaide University, Maxine Shephard used 2AFC procedure with a four-point rating scale to measure the discriminability of serial positions in a five-item memory span task. After presentation of five different digits in random order followed by a rehearsal preventing task, observers were presented with two items which had occurred in the sequence and asked to judge whether the order of the test items was the same as that of the original stimulus

176

items. The ROC curves, as predicted by detection theory, were straight lines when plotted on a double-probability scale, and had slopes of about 1. While both the Schulman & Mitchell and Markowitz & Swets studies yielded ROC curves which were straight lines when plotted on a double-probability scale, the latter study showed that slopes decreased as signal strength increased. This finding runs counter to the prediction of detection theory that 2AFC ROC curves should have unit slopes. Nonetheless, experimental data support the detection theory formulation of the 2AFC task more than they support the views of high threshold theory.

mAFC tasks and high threshold theory

High threshold theory's formulation of the mAFC task follows easily from consideration of the 2AFC task. From formulae (7.9) and (7.14), a stimulus-response matrix for the high threshold theory version of the 2AFC task can be drawn up showing the probability of different stimulus-response outcomes in terms of p, the proportion of true correct responses, and g, the guessing factor. This matrix is shown in Table 7.3, which should be compared with Table 6.1, the stimulus-response matrix for choice theory.

TABLE 7.3 *The stimulus-response matrix for the 2AFC task in terms of p, the true proportion of correct responses, and g, the guessing factor, according to high threshold theory*

		Response event	
		$\langle SN \rangle$	$\langle NS \rangle$
Stimulus event	$\langle sn \rangle$	$P(\langle SN \rangle \mid \langle sn \rangle)$ $= p + g(1 - p)$	$P(\langle NS \rangle \mid \langle sn \rangle)$ $= (1 - p)(1 - g)$
	$\langle ns \rangle$	$P(\langle SN \rangle \mid \langle ns \rangle)$ $= g(1 - p)$	$P(\langle NS \rangle \mid \langle ns \rangle)$ $= p + (1 - p)(1 - g)$

It is easy to extend this matrix to any number of alternatives, and as an example, the matrix for a 3AFC task is shown in Table 7.4. Compare it with Table 6.3, the choice theory version of the same matrix.

A special case of the mAFC task can now be considered. This is when all items in the mAFC task can be considered to have the same

177

TABLE 7.4   *The stimulus-response matrix for the 3AFC task according to high threshold theory*

		Response alternative		
		1	2	3
	1	$p_1 + g_1(1 - p_1)$	$g_2(1 - p_1)$	$g_3(1 - p_1)$
Stimulus alternative	2	$g_1(1 - p_2)$	$p_2 + g_2(1 - p_2)$	$g_3(1 - p_2)$
	3	$g_1(1 - p_3)$	$g_2(1 - p_3)$	$p_3 + g_3(1 - p_3)$

sensitivities and bias, i.e. when p and g are the same for all items. The stimulus-response matrix for this case is shown in Table 7.5.

All rows in this matrix are the same, and all rows sum to 1. For row 1 of Table 7.5 it can be seen that the row sum is the probability of giving response 1 to stimulus 1, plus the probability of giving

TABLE 7.5   *The stimulus-response matrix according to high threshold theory for an mAFC task in which sensitivities and bias for all items are the same*

		Response alternative				
		1	2	3 …	m	Row total
	1	$p + g(1 - p)$	$g(1 - p)$	$g(1 - p)$	$g(1 - p)$	1·0
	2	$g(1 - p)$	$p + g(1 - p)$	$g(1 - p)$	$g(1 - p)$	1·0
Stimulus Alternative	3	$g(1 - p)$	$g(1 - p)$	$p + g(1 - p)$	$g(1 - p)$	1·0
	⋮					
	m	$g(1 - p)$	$g(1 - p)$	$g(1 - p)$	$p + g(1 - p)$	1·0

response 2 to stimulus 1, etc. The probability of giving response 1 to stimulus 1, which is the correct response, is $p + g(1 - p)$. There are also $m - 1$ incorrect responses, each of which has a probability of $g(1 - p)$. So

$$p + g(1 - p) + (m - 1)g(1 - p) = 1. \qquad (7.15)$$

If (7.15) is rearranged to find g in terms of p and m, p disappears and we are left with

$$g = 1/m.$$

The probability, $P(c)$, of making a correct response in the mAFC

178

task is the same as the probability of giving response 1 to stimulus 1, so that

$$P(c) = p + g(1 - p),\qquad(7.16(a))$$

and by substituting $1/m$ for g in (7.16(a)) we obtain

$$P(c) = p + \frac{1}{m}(1 - p).\qquad(7.16(b))$$

Formula (7.16(b)) is the familiar 'correction for chance guessing' for an mAFC task. It states that the observed proportion of correct responses in such a task, $P(c)$, consists of a proportion of true correct responses, p, plus a proportion of lucky guesses, $(1 - p)/m$. If the number of alternatives is increased in an mAFC task, p should remain constant and $P(c)$ should decrease. This follows from (7.16(b)). Signal detection theory also predicts that $P(c)$ should decline as m increases, but for different reasons, as was seen in Chapter 3. Experiments using mAFC tasks, and in which the number of items used in the forced-choice task were varied, have generally reported that $P(c)$ does fall as m is increased. The best known of these experiments is a word-recognition test reported by Miller, Heise & Lichten (1951) in which observers attempted to recognize monosyllables masked by white noise. In this experiment both the signal-to-noise ratio and the number of alternatives in the forced-choice recognition task were varied from 2 to 256. Murdock (1963) obtained similar findings in forced-choice responses to a general information questionnaire where the number of choices was 2, 3 or 4. Teghtsoonian (1965), using 2, 4, 8 or 16 alternatives in a test of short-term memory, also found that $P(c)$ declined as m increased. These are a few typical examples of the relationship between $P(c)$ and m.

Both high threshold theory and signal detection theory predict that the observed proportion of correct responses should decline as the number of alternatives increases. but it is possible to distinguish between them by seeing whether it is p or d' which remains the more invariant as m is changed. If (7.16(b)) is used to find p from values of $P(c)$ obtained in a number of forced-choice tasks it is found that p itself declines as m increases (this certainly applies to the Miller, Heise & Lichten data). On the other hand, if Elliott's (1964) tables are used to calculate d' from $P(c)$, this measure of sensitivity remains

179

relatively stable with changes in m (Green & Birdsall, 1964; Swets, 1959). High threshold theory is certainly not a credible model for the mAFC task, and although detection theory is, there are other models which adequately fit the Miller, Heise & Lichten data. Some of these will be described briefly.

(a) *Sophisticated guessing models* (Broadbent, 1967; Murdock, 1963; Stowe, Harris & Hampton, 1963). These assume that although the observer in an mAFC task may fail, on some trials, to obtain sensory evidence which unambiguously allows him to choose the correct item, he may at least be able to eliminate a proportion of incorrect items. As in the simple guessing model, he must still make a random choice of his response, but the choice is from the subset of items remaining after some of the m alternatives have been eliminated as certainly incorrect. The version of this model proposed by Stowe, Harris & Hampton (1963) fits the Miller, Heise & Lichten data quite well.

(b) *Choice Theory*. Luce's choice theory, encountered in Chapter 6, also fits the mAFC data of Miller, Heise & Lichten (Luce, 1963) (see Problems 4 and 5 at the end of this chapter for more information about the choice theory approach to the mAFC task).

(c) *Information Theory*. Garner (1962, Figure 3.5) has fitted the Miller, Heise & Lichten data to an information theory model which predicts that the amount of information transmitted should remain constant for changes in the number of alternatives for a given signal-to-noise ratio. Again, the data fit his model tolerably well.

The fact remains, that although the best data currently available will not discriminate between a number of accounts of the mAFC task, signal detection theory does at least as well as its rivals, and much better than a simple guessing model based on high threshold theory.

OTHER EVIDENCE FAVOURING DETECTION THEORY

Detection theory may be evaluated in a number of other ways besides those discussed in detail above. For instance, it is desirable that for given levels of signal and noise, the sensitivity measure should

180

remain invariant with changes in experimental procedure. High threshold theory fails in this respect as Blackwell (1953) has reported that forced-choice procedures yield lower estimates of the threshold than do yes–no procedures. By contrast, Egan, Schulman & Greenberg (1959) obtained comparable values of Δm when using a four-point rating scale, or a series of three separate yes–no tasks to obtain points for ROC curves. Swets (1959) obtained similar values of d' by using yes–no, 2AFC, or 4AFC procedures.

Another finding which supports the notion that detection theory adequately describes perceptual decision processes is that second choices in mAFC tasks show better than chance discrimination. Swets, Tanner & Birdsall (1961) conducted a 4AFC task in which. if an observer selected an incorrect alternative when responding, he was asked to make a second choice. According to high threshold theory, if the observer cannot select the correct item on his first attempt, all subsequent attempts should be at random. So for the 4AFC task, the probability of selecting the correct item on the second attempt, given that the first attempt has failed, should be $\frac{1}{3}$. According to the detection theory description of the mAFC task given in Chapter 3, the observer will select as his first choice the item whose value of x is the greatest. If it turns out that this item is an incorrect one, it is still most likely that the item with the second greatest value of x will be the correct one, and will be chosen on the second attempt. The results of the experiment showed that second choices were better than chance guessing, thus supporting detection theory in preference to high threshold theory. Similarly Brown (1964, 1965) and McNicol (1971) have found that second choices in learning and memory experiments are not random, but preserve information about the correct response.

One final example of the adequacy of a model based on detection theory rather than on high threshold theory is Broadbent's (1967) study of the word frequency effect. This is the tendency for observers in perceptual and learning tasks where they are presented with stimuli which are either common or uncommon words, to perceive or recall the common words more accurately than the uncommon words. Considered in terms of high threshold theory the word frequency effect might be attributed to observers having a stronger tendency to guess familiar words for their responses when the stimulus word fell below the threshold. The observed proportion

181

of correct responses to common stimuli would thus consist of a proportion of correct responses .plus a substantial proportion of lucky guesses. The observed proportion of correct responses to uncommon stimuli would also consist of a proportion of true correct responses, but these would be supplemented by few lucky guesses, as observers would rarely use uncommon words as responses when guessing. Broadbent's experiments showed that the word frequency effect could not be accounted for by a guessing model of the type implied by high threshold theory, but was compatible with the type of response bias predicted by signal detection theory.

Having said all of this, it would be incorrect to leave the impression that signal detection theory, in the form in which it has been presented here, is the sole candidate for explaining how perceptual decisions are made. If a high threshold looks an unpromising candidate then a low threshold might be considered in its place. That is to say, there is a value of x, set well within the noise distribution, below which stimuli will have no sensory effect. One version of this proposed by Swets, Tanner & Birdsall (see Swets, 1961) proposes that the ROC curve is bow-shaped but with a linear segment at the bottom. That is, while observers do not discriminate among values of x lying below the threshold, they can discriminate among the values lying above the threshold in the manner normally proposed by signal detection theory. Another version proposed by Luce (1959, 1963) assumes that observers neither discriminate between values of x lying below the low threshold, nor between values lying above the threshold. The ROC curve for this theory consists of two straight line segments, one starting at $(0, 0)$, the other at $(1, 1)$ and meeting so as to produce a dog-leg curve not easily discriminated from the bow-shaped curve of signal detection theory. If one threshold is not enough then multi-threshold models can be introduced. Green (see Swets, 1961) has proposed a two-threshold model whose ROC curve consists of three linear segments. Such a theory copes as well with the available experimental data as does detection theory. If data arose which were outside the range of two-threshold theory we could easily fall back on a three-threshold theory, and so on. Of course the more thresholds. the more the theory approximates detection theory so that there will always be some multi-threshold theory which can account for the results encompassed by detection theory. Despite the fact that there are other possibilities besides detection theory, none of them are actually better in dealing with experimental data obtained to date, and many are so close to detection theory in their predictions that it would be unlikely that experiments will distinguish between them and detection theory.

Problems

1. From the table below, which is raw data from a rating scale experiment, determine the *a posteriori* probabilities associated with each confidence level.

		Response event Certain signal to certain noise				
		1	2	3	4	5
Stimulus event	s	426	194	127	60	33
	n	106	196	239	171	128

2. In an experiment to determine the absolute threshold for audibility of a pure tone, observers are presented with the tone several times at a number of different intensities and asked to say if they can hear it. The data in the table below show the proportion of occasions on which observers correctly detected the tone at each intensity level. (Intensity is scaled in an arbitrary unit which may range from 0 to 1). From the data find the average value of the absolute threshold, $\bar{L}$ in intensity units.

(*Hint*: It is easier to interpolate on straight lines than it is on curved ones, and a normal ogive plotted as z-values is a straight line.)

If blank trials had been inserted amongst the stimuli, and if observers had been trained to keep their false alarm rates constant at $P(S|n) = 0.01$ for all intensity levels, find d' for each intensity level.

	Intensity units					
Probability of a correct detection	0·40	0·50	0·55	0·65	0·70	0·80
	0·02	0·16	0·31	0·69	0·84	0·98

183

N

3. In a 2AFC task in which confidence ratings were used, the raw data in the table below were obtained. Estimate p, the proportion of true correct responses for each pair of $P(\langle SN \rangle | \langle sn \rangle)$ and $P(\langle SN \rangle | \langle ns \rangle)$ values and compare these values with that of d'_{FC} as estimated from the 2AFC ROC curve.

| | | Observer's response High confidence $\langle SN \rangle$ to high confidence $\langle NS \rangle$ | | | |
		1	2	3	4
Stimulus event	sn	92	72	20	16
	ns	14	48	38	100

4. Table 7.5 is the stimulus-response matrix for high threshold theory which assumes equal sensitivities and biases for all m items in a mAFC task, and it is from this matrix that formula (7.16) was obtained. Construct a similar matrix for choice theory assuming equal sensitivities and biases on all items, and deduce the choice theory formula for the mAFC task.

5. In the table below are several values of $P(c)$ obtained from mAFC tasks with the same signal-to-noise ratios, but with different values of m. Elliott's (1964) forced-choice tables show that each value of $P(c)$ gives the same d' value of 1, i.e. according to signal detection theory all $P(c)$ values represent equivalent degrees of sensitivity.

Do the $P(c)$s represent equivalent degrees of sensitivity for high threshold theory and for choice theory? If not, calculate the sensitivity measures for each $P(c)$ for the two theories. (Use the mAFC formula for choice theory which you worked out in problem 5).

| | Number of alternatives in mAFC task | | | | | |
	2	4	8	16	32	256
$P(c)$	0·76	0·56	0·38	0·26	0·16	0·04

Chapter 8

THE LAWS OF CATEGORICAL AND COMPARATIVE JUDGEMENT

ANTECEDENTS OF SIGNAL DETECTION THEORY

High threshold theory, although a manifestly inadequate account of decision processes in perception and memory, has been discussed at length because of the powerful sway it has held over the thinking and methodology of experimental psychologists for the last century (Corso, 1963). Psychologists have attempted to measure absolute and difference thresholds, studies of verbal learning have employed corrections for chance guessing, and learning theories (e.g. Hull, 1943) have embodied threshold concepts. Many data are thrown into doubt if the assumptions of signal detection theory are true, although the results of some experiments based on the methodology of threshold theory may be translated into detection theory measures (Treisman & Watts, 1966).

Nevertheless, if the majority of experimental psychologists preferred to think of thresholds and guessing corrections, it was not because models like detection theory had not been thought of. Thurstone (1927a) proposed first the Law of Categorical Judgement, which parallels yes–no rating tasks in many respects. and second. (1927b) the Law of Comparative Judgement, which is an analysis of the 2AFC task. Basic to both these laws is the notion that sensory evidence is variable from moment to moment, and that for a particular stimulus, the distribution of sensory effect is continuous and Gaussian. Thurstone called these distributions of sensory effect *discriminal processes*, and they are quite analogous to the distributions of signal and noise of signal detection theory. He was interested in determining the *category boundaries* (or criteria) and the *discriminal dispersions* (or standard deviations) of the underlying

185

distributions, and adopted the method, well-known to us now, of obtaining sets of cumulative probabilities and converting them into z-scores. This procedure was called the method of successive intervals, and Saffir (1937) describes how one set of z-scores may be plotted against another to give the straight-line function which we now know as the ROC curve. Torgerson (1958) describes the methods of Thurstonian scaling, and Lee (1969) gives an illuminating series of comparisons between the methods proposed by Thurstone and those of signal detection theory.

Although it is interesting to know that much of what we thought was new was really old, and no doubt there are morals which can be drawn from the insistence of many psychologists on using bad psychophysical procedures when good ones were available, Thurstonian scaling has other things to offer in addition to the lessons of history. With its assistance, signal detection theory can be extended to look at some rather interesting problems. We begin first by considering the Law of Categorical Judgement.

CATEGORY RATING TASKS
(Torgerson 1958, Ch. 10)

Thurstonian scaling for the method of successive intervals

Consider an experiment in which observers have to judge the loudness of a short burst of white noise. Galanter & Messick (1961) used twenty different sound pressure levels and presented these several times each in random order. After each presentation, observers were asked to indicate the loudness of the stimulus on an eleven-point scale, using 'one' for the softest stimulus and '11' for the loudest. We will consider a hypothetical example in which four stimulus and response categories have been used. The results of an observer can be represented in a matrix of 4 stimuli × 4 responses called a *confusion matrix*, an example of which is given in Table 8.1.

Just as in signal detection theory, we can consider the sensory effect of each stimulus being due to the effects of the stimulus plus the effects of internal noise. We would therefore expect the distribution of sensory effect for a stimulus to be Gaussian, due to the variability of the internal noise, and to have a mean value, d', the

TABLE 8.1 *Confusion matrix for a category rating task in which the observer is presented with 4 stimulus events and allocates them to 4 response categories*

		Response category				
		R_1	R_2	R_3	R_4	Row sum
	S_1	50	34	14	2	100
Stimulus	S_2	40	20	17	23	100
event	S_3	31	19	19	31	100
	S_4	2	14	34	50	100

size of which would be determined by the strength of the stimulus. The distributions of sensory effect for the four stimuli in our example could therefore be represented in a manner shown in Figure 8.1.

The mean of S_1 is arbitrarily fixed at 0, as is done for the noise distribution mean in ordinary signal detection analysis, and all other distances are measured relative to this mean. The observer

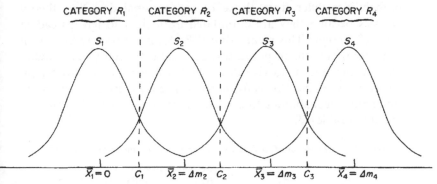

FIGURE 8.1 *Distributions of sensory effect for four stimuli, S_1, S_2, S_3, S_4, in a category rating task. Response categories R_1, R_2, R_3, R_4 are determined by placemat of criteria C_1, C_2, C_3*

establishes three criteria, C_1, C_2, C_3, and if the value of x for a stimulus falls below C_1, he assigns it to response category 1; if it falls between C_1 and C_2, he assigns it to response category 2, and so on. As the distributions of sensory effect are Gaussian, it will turn out that S_1 will sometimes produce a value of x which will be misclassified into categories 2, 3 or 4, and similar errors will occur when responding to the other three stimuli.

187

TABLE 8.2 *Sets of conditional probabilities. $P(R_j|S_i)$, for the category rating task whose raw data are shown in Table 8.1*

| | | \multicolumn{4}{c}{Response category} | | | |
		R_1	R_2	R_3	R_4
	$P(R_j|S_1)$	1·00	0·50	0·16	0·02
Conditional	$P(R_j|S_2)$	1·00	0·60	0·40	0·23
probability	$P(R_j|S_3)$	1·00	0·69	0·50	0·31
	$P(R_j|S_4)$	1·00	0·98	0·84	0·50

Table 8.1 can thus be thought of in an analogous fashion to the raw data matrices obtained from rating scale tasks, except that instead of two rows, one for responses to signal, the other for responses to noise, we now have four rows, each corresponding to a different stimulus. However we can still go ahead and convert these into sets of cumulative probabilities, as has been done in Table 8.2. The probabilities are values of $P(R_j|S_i)$. The cumulative probabilities differ from those of conventional detection theory analysis because they are cumulated from right to left instead of from left to right. This means that $P(R_j|S_i)$ is the probability, given stimulus S_i, that the response was assigned to category j or a higher one. All this does is to give Δm values with a positive rather than a negative sign, but this is for convenience only. By converting the cumulative probabilities to z-scores, a matrix of $z(R_j|S_i)$ values is obtained (Table 8.3).

TABLE 8.3 *Conversion of the $P(R_j|S_i)$ values in Table 8.2 into values of $z(R_j|S_i)$*

| | \multicolumn{4}{c}{Response category} | | | |
	R_1	R_2	R_3	R_4
$z(R_j|S_1)$	—	0·00	+1·00	+2·00
$z(R_j|S_2)$	—	−0·25	+0·25	+0·75
$z(R_j|S_3)$	—	−0·50	0·00	+0·50
$z(R_j|S_4)$	—	−2·00	−1·00	0·00

Just as a similar set of z-scores enabled us to construct ROC curves for signal an noise, from which the characteristics of the underlying distributions could be inferred, so also can Table 8.3 be used to obtain ROC curves from which the means and standard deviations of the distributions for the four stimuli, and the positions of the three criteria, can be determined. In a similar manner to the noise distribution of conventional signal detection analysis, one of the four stimulus distributions must have its mean fixed at 0 and its standard deviations at 1, so that the means and standard deviations of all other distributions can be measured relative to it. The distribution for S_1 will be used as this reference point, but the choice is quite arbitrary.

By plotting values of $z(R_j | S_2)$ against those of $z(R_j | S_1)$ an ROC curve is obtained from which the value of $\Delta m_{1,2}$ (the value of $z(R_j | S_1)$ when $z(R_j | R_2) = 0$) and the slope of the ROC curve can be

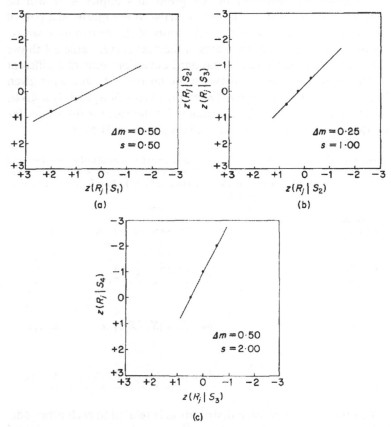

FIGURE 8.2   ROC *curves constructed from the data of Tables 8.3*

189

obtained. $\Delta m_{1,2}$ is the distance between the means of the distributions for S_1 and S_2. The slope is equal to the ratio σ_1/σ_2, and as σ_1 has been fixed at 1, the value of σ_2 is the reciprocal of the slope. The ROC curve is shown in Figure 8.2(a).

Likewise the ROC curve of $z(R_j|S_3)$ against $z(R_j|S_2)$ (Figure 8.2(b)) will give the distance between the means of S_2 and S_3, and a slope equal to the ratio σ_2/σ_3. Finally, by plotting $z(R_j|S_4)$ against $z(R_j|S_3)$ a third ROC curve is obtained whose Δm value is the distance between the means of S_3 and S_4, and whose slope is the ratio σ_3/σ_4 (Figure 8.2(c)).

Recalling the definition of Δm given in Chapter 4, it will be remembered that it is the difference between the means of a pair of normal distributions, scaled in S.D. units of the distribution whose z-values are plotted on the x-axis of the ROC curve. Table 8.4 shows that each Δm is expressed in standard deviation units of a different distribution. The distance between the means of S_1 and S_2 is given in S.D.s of S_1, the distance between the means of S_2 and S_3 is given in S.D. units of S_2, etc. The formulae for the various Δm values for a category rating task with n stimuli are shown in Table 8.4.

TABLE 8.4 *Formulae for Δms and slopes of ROC curves based on distributions of sensory effect for a rating category task with n stimuli. The ROC curves have been formed by plotting $z(R_j|S_{i+1})$ against $z(R_j|S_i)$ where S_i is the ith stimulus and S_{i+1} is the stimulus next to it*

Variables for ROC curve		Distance between means (Δm)	Slope (s)
x-variable	y-variable		
S_1	S_2	$\Delta m_{1,2} = (\bar{X}_2 - \bar{X}_1)/\sigma_1$	σ_1/σ_2
S_2	S_3	$\Delta m_{2,3} = (\bar{X}_3 - \bar{X}_2)/\sigma_2$	σ_2/σ_3
S_3	S_4	$\Delta m_{3,4} = (\bar{X}_4 - \bar{X}_3)/\sigma_3$	σ_3/σ_4
.			
.			
S_i	S_{i+1}	$\Delta m_{i,i+1} = (\bar{X}_i - \bar{X}_{i+1})/\sigma_i$	σ_i/σ_{i+1}
.			
.			
S_{n-1}	S_n	$\Delta m_{n-1,n} = (\bar{X}_{n-1} - \bar{X}_n)/\sigma_{n-1}$	σ_{n-1}/σ_n

To understand how each distribution is related to each other one, it is necessary to scale all these distances in a common unit of measurement. The distribution of S_1 has been assumed as this reference point. with its mean fixed at 0 and its S.D. at 1. The Δm values for all other distribution means must be expressed in σ_1 units.

The slopes of the ROC curves can first be used to find the standard deviations of each distribution. The first ROC curve in Figure 8.2 is a plot of $z(R_j \,|\, S_2)$ against $z(R_j \,|\, S_1)$ and from Table 8.4 it can be seen that its slope $s_{1,2}$ is the ratio σ_1/σ_2. As $\sigma_1 = 1$, then σ_2 is the reciprocal of the slope, which is 2·00. The second ROC curve in Figure 8.2 is a plot of $z(R_j \,|\, S_3)$ against $z(R_j \,|\, S_2)$, and from Table 8.4 its slope, $s_{2,3}$, is σ_2/σ_3. As σ_2 is already known to be 2·00, then σ_3 can be determined, and is equal to 2·00 also. Finally, in the third ROC curve of Figure 8.2 we have a plot of $z(R_j \,|\, S_4)$ against $z(R_j \,|\, S_3)$, whose slope, $s_{3,4}$, is equal to σ_3/σ_4. σ_3 has been found to be 2·00 so that σ_4 is 1·00. A general set of equations for finding the standard deviations of a set of distributions from the slopes of the ROC curves can be written, and these are shown in Table 8.5. The standard deviations for the example data whose ROC curves are shown in

TABLE 8.5 *General formulae for calculating standard deviations of distributions of sensory effect for stimuli in a category rating task by use of the slopes of the ROC curves.*

Stimulus	Formula for calculating S.D. from ROC curve slopes
S_1	$\sigma_1 = 1$
S_2	$\sigma_2 = \dfrac{1}{s_{1,2}}$
S_3	$\sigma_3 = \dfrac{1}{s_{1,2}}\dfrac{1}{s_{2,3}}$
S_4	$\sigma_4 = \dfrac{1}{s_{1,2}}\dfrac{1}{s_{2,3}}\dfrac{1}{s_{3,4}}$
$\vdots$	
S_i	$\sigma_i = \dfrac{1}{s_{1,2}}\dfrac{1}{s_{2,3}}\dfrac{1}{s_{3,4}}\ldots\dfrac{1}{s_{i-1,i}}$
$\vdots$	
S_n	$\sigma_n = \dfrac{1}{s_{1,2}}\dfrac{1}{s_{2,3}}\dfrac{1}{s_{3,4}}\ldots\dfrac{1}{s_{i-1,i}}\ldots\dfrac{1}{s_{n-1,n}}$

Figure 8.2 are calculated in Table 8.6 with the aid of these equations.

Having found the standard deviations, the next step is to rescale all the Δm values in units of σ_1. Referring to Table 8.4 it can be seen

TABLE 8.6. *Calculation of standard deviations for the four stimuli whose* ROC *curves are shown in Figure 8.2.*

Stimulus	Calculation of S.D. from ROC curve slopes
S_1	$\sigma_1 = 1 \cdot 00$ (by definition)
S_2	$\sigma_2 = \dfrac{1}{0 \cdot 50} = 2 \cdot 00$
S_3	$\sigma_3 = \dfrac{1}{0 \cdot 50} \dfrac{1}{1 \cdot 00} = 2 \cdot 00$
S_4	$\sigma_4 = \dfrac{1}{0 \cdot 50} \dfrac{1}{1 \cdot 00} \dfrac{1}{2 \cdot 00} = 1 \cdot 00$

that $\Delta m_{1,2}$ is already scaled in the appropriate unit. $\Delta m_{2,3}$ is scaled in units of σ_2 and to convert it to σ_1 units the following transformation needs to be used:

$$\Delta m'_{2,3} = \frac{(\overline{X}_3 - \overline{X}_2)}{\sigma_2} \cdot \frac{\sigma_2}{\sigma_1}, \qquad (8.1(a))$$

where $\Delta m'_{2,3}$ is the rescaled value of $\Delta m_{2,3}$, σ_1 has been defined as 1, and $(\overline{X}_3 - \overline{X}_2)/\sigma_2$ is the expression for the original value of Δm. The formula for the rescaled value of $\Delta m'_{2,3}$ could also be written as

$$\Delta m'_{2,3} = \Delta m_{2,3} \cdot \sigma_2. \qquad (8.1(b))$$

In a similar manner $\Delta m'_{3,4}$, the rescaled value of $\Delta m_{3,4}$ is found from the expression

$$\Delta m'_{3,4} = \frac{(\overline{X}_4 - \overline{X}_3)}{\sigma_3} \cdot \frac{\sigma_3}{\sigma_1}, \qquad (8.2(a))$$

which may be rewritten as

$$\Delta m'_{3,4} = \Delta m_{3,4} \cdot \sigma_3. \qquad (8.2(b))$$

In general, then:

$$\Delta m'_{i,i+1} = \frac{(\bar{X}_{i+1} - \bar{X}_i)}{\sigma_i} \cdot \frac{\sigma_i}{\sigma_1} \qquad (8.3(a))$$

which may be rewritten as

$$\Delta m'_{i,i+1} = \Delta m_{i,i+1} \cdot \sigma_i. \qquad (8.3(b))$$

By using the fomulae for rescaling Δm values in conjunction with the standard deviations in Table 8.6, values of $\Delta m'$ for each stimulus may be obtained. These are shown in Table 8.7.

TABLE 8.7   *Rescaling of Δm values from the* ROC *curves in Figure 8.2 in S.D. units of the distribution for S_1*

Distance between stimulus means	Δm from ROC curve	Multiplier for rescaling in σ_1 units	Rescaled value of $\Delta m'$
$\bar{X}_2 - \bar{X}_1$	0·50	$\sigma_1 = 1\cdot00$	$0\cdot50 \times 1\cdot00 = 0\cdot50$
$\bar{X}_3 - \bar{X}_2$	0·25	$\sigma_2 = 2\cdot00$	$0\cdot25 \times 2\cdot00 = 0\cdot50$
$\bar{X}_3 - \bar{X}_4$	0·50	$\sigma_3 = 2\cdot00$	$0\cdot50 \times 2\cdot00 = 1\cdot00$

At this point the means and standard deviations of all the distributions have been expressed in a common standard deviation unit, so that they may be directly compared with each other. The same thing could be done with the criterion points as well. The $z(R_j|S_1)$ values in Table 8.3 already represent the distances of each criterion from the mean of S_1 in S.D. units of σ_1, and it might be thought that this is an adequate means of estimating their positions. However in the example, errorless data have been used. When there has been error in estimating the $z(R_j|S_i)$ values, it would be wiser not to rely solely on the $z(R_j|S_1)$ values alone, but to obtain additional estimates of the positions of the criteria from the other rows in the confusion matrix. This will necessitate rescaling the $z(R_j|S_i)$s so that they are expressed in S.D. units of σ_1 in the same manner as Δm values were rescaled. Also, as each $z(R_j|S_i)$ is the distance of a criterion from the mean of distribution S_i, it will be also necessary to express the distance as one from the mean of S_1. After this has been done, the several estimates of each criterion from the mean of S_1

193

may be averaged to obtain a best estimate of each criterion's position.

The method which has been described for Thurstonian scaling is not necessarily the best one to use with real data, but it is the most meaningful for the purposes of explanation. Torgerson (1958) describes a number of the traditional methods for dealing with data from a category rating task, and since that time, Schönemann & Tucker (1967) have presented a maximum likelihood solution for dealing with this type of data.

Rescaling signal and noise distributions

The discussion of Thurstonian scaling in successive interval tasks has shown that methods quite analogous to those of signal detection theory can be used to examine distributions of sensory effect for a range of stimuli, and need not be restricted to simple comparisons between pairs of distributions, as is normally done in yes–no and rating-scale experiments. By the same token, the method opens the way to simultaneously comparing several signal and noise distributions whose means and standard deviations have been obtained by conventional detection theory techniques. This extension of signal detection theory was suggested first by Ingleby (1968) who demonstrated its usefulness in a series of experiments on perception and memory. Wickelgren (1968) has proposed a uni-dimensional strength theory which also integrates the concepts of signal detection theory and those of Thurstone's Laws of Categorical and Comparative Judgement. Although Ingleby's method for rescaling the distributions differs from the one described in the previous section, they are essentially his ideas which will be illustrated in the following paragraphs.

Consider the following detection task conducted by McNicol & Willson (1971). Observers were shown a visual display for a few milliseconds which comprised a string of six typed capital letters. They were asked to decide, after each presentation, whether the string contained the target letter 'Q', which had an 0·5 chance of appearing on any trial, and whose position in the six-letter string was varied at random from trial to trial. The decision was made in the form of a confidence judgement on a five-point scale. Unknown to the observers, two types of non-target letters were used. Sometimes they were the letters C, D, G, O, S, U, which resemble the shape of the target. On other occasions they were the letters I, T, V, X, Y, Z, whose shapes are dissimilar to that of the target. Thus, at random, the observer would be shown a letter string in which all the characters were rather similar to the target he was trying to detect, or he would be shown a string in which the non-targets were markedly different from the target. In either case he had to decide whether the target was amongst them.

The raw data from the experiment consist of responses to signal and responses to noise for detection of the target from similar non-targets, and responses to signal and noise for detection of the target amongst dissimilar non-targets. These raw data can be converted into sets of $P(S|s)$ and $P(S|n)$ values for the two experimental conditions. The groups results are shown in Table 8.8.

TABLE 8.8 *Hit and false alarm rates for the two experimental conditions of the letter recognition task conducted by McNicol and Willson*

(a) Detection of target amongst dissimilar non-targets.
Observer's response
High certainty signal to high certainty noise

	1	2	3	4	5
$P(S|s)$	0·59	0·75	0·86	0·93	1·00
$P(S|n)$	0·02	0·05	0·13	0·32	1·00

(b) Detection of target amongst similar non-targets
Observer's response
High certainty signal to high certainty noise

	1	2	3	4	5
$P(S|s)$	0·51	0·74	0·89	0·96	1·00
$P(S|n)$	0·13	0·36	0·64	0·85	1·00

These $P(S|s)$ and $P(S|n)$ values are converted into $z(S|s)$ and $z(S|n)$ values in the usual fashion, and appear in Table 8.9. By plotting $z(S|s)$ against $z(S|n)$ for detection of the target amongst both similar and dissimilar non-targets, two ROC curves are obtained (Figure 8.3).

By obtaining the values of Δm and σ_s from these curves it is possible to reconstruct the underlying distributions of signal and noise for the two experimental conditions. These are shown in Figure 8.4 and it has been assumed, as is conventional in signal detection analysis, that each noise distribution has a mean of 0 and an S.D.

195

TABLE 8.9 *Conversion of the $P(S \mid s)$ and $P(S \mid n)$ values for the letter recognition task data in Table 8.8 into $z(S \mid s)$ and $z(S \mid n)$ values*

(a) Detection of target amongst dissimilar non-targets

| | Observer's response High certainty signal to high certainty noise | | | |
	1	2	3	4
$z(S \mid s)$	−0·23	−0·68	−1·08	−1·48
$z(S \mid n)$	+2·05	+1·65	+1·13	+0·47

(b) Detection of target amongst similar non-targets

| | Observer's response High certainty signal to high certainty noise | | | |
	1	2	3	4
$z(S \mid s)$	−0·03	−0·64	−1·23	−1·75
$z(S \mid n)$	+1·13	+0·34	−0·34	−1·04

of 1. The impression gained from Figure 8.4 is that sensitivity in detecting the target is better when it is amongst dissimilar non-targets, as the signal distribution for the dissimilar condition is further up the x-axis than the signal distribution for the similar condition. Both signal distributions appear to have the same standard deviation, and the criteria for the dissimilar condition are more closely bunched together and further up the x-axis than the criteria for the similar condition.

Now Thurstonian scaling techniques will be applied to the same data. As a start, one of the distributions, in this case the noise distribution in the similar condition, will be selected as a reference point. Its mean will be fixed at 0 and its S.D. at 1. The other three distributions, noise for the dissimilar condition, and signal for both similar and dissimilar conditions, will be scaled in S.D. units of this first noise distribution. In the example for the category rating task given earlier in this chapter, the rescaling procedure involved plotting the $z(R_j \mid S_i)$ values for one stimulus against those for an adjacent stimulus. The reason for doing this was to obtain ROC curves from distributions with as much overlap as possible, because these would be likely to yield the greatest number of points for the ROC curves, and hence to give the most reliable estimates of Δm and s. In this example we will adopt the cruder (and not normally recom-

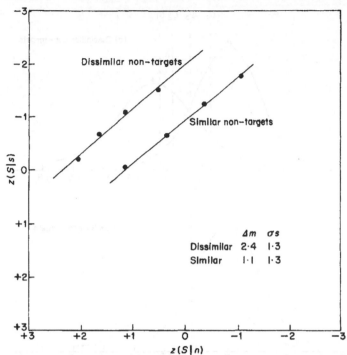

FIGURE 8.3 ROC *curves for the two conditions in the letter recognition task with each signal distribution sealed against its noise distribution (McNicol and Willson 1971)*

mended) procedure of plotting z-values for noise in the dissimilar condition, and for both types of signal, against $z(S \mid n)$ for the similar condition. Three ROC curves will be obtained in which the three other distributions are scaled against the noise distribution being used as a reference point. The $\Delta m'$ values obtained from these ROC curves will be the distances of each distribution from the mean (of 0) of the noise distribution for the similar condition, and the reciprocals of the slopes will be the S.D.s of the three distributions relative to the S.D. (of 1) of the reference distribution. In Figure 8.5 the z-scores from Table 8.9 are plotted in this fashion.

From the $\Delta m'$ and σ values obtained from Figure 8.5, another picture of the distributions of signal and noise for the two experimental conditions may be obtained (Figure 8.6). This time they are scaled on a common x-axis, and it is assumed that the positions of

197

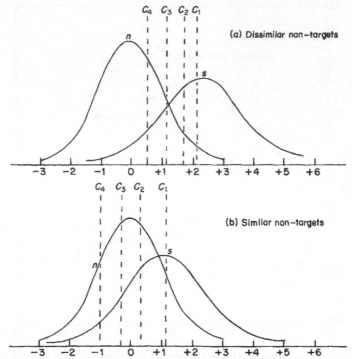

FIGURE 8.4 *The distributions of sensory effect for the two conditions in the letter recognition task. Each noise distribution has been fixed with a mean of O and an S.D. of 1*

the criteria are not changed from one condition to the other. This assumption seems reasonable in this case, as trials from one condition were randomly mixed with trials from the other, and observers had no way of knowing before a trial was given, whether they would have to discriminate the target from similar or dissimilar non-targets.

The impression gained from Figure 8.6 is different from that given by Figure 8.5. The noise distribution for the dissimilar condition lies well below the noise distribution for the similar condition, but both signal distributions lie about the same distance above the reference distribution. In addition, it appears that the standard deviations of the signal and noise distributions in the dissimilar condition are greater than those for the similar condition.

198

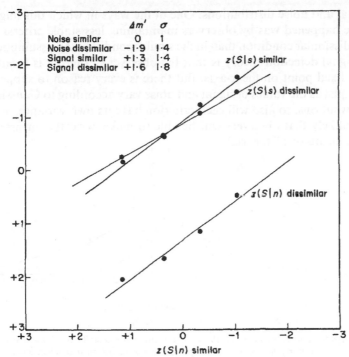

FIGURE 8.5   ROC *curves obtained by plotting all other sets of z-values from Table 8.9 against z(S/n) for dissimilar non-targets*

The major effect of changing similar non-targets into dissimilar non-targets is therefore to lower the level of noise. The distribution for the signal, which itself remains the same in both conditions, is left virtually unaltered. The idea that one type of noise may have different consequences from another is a useful innovation. In experiments where truly Gaussian white noise is used, it is reasonable to assume that the noise distribution always remains fixed and it is only the signal distribution which varies. Experiments which involve more complex stimulus situations may, however, wish to examine changes in noise as well as changes in signal, and the Thurstonian scaling techniques are suited to this type of analysis.

Criterion variability

One other effect which seems to have been caused by changing similar to dissimilar non-targets was to increase the variances of the

199

o

signal and noise distributions. One of the ways in which this might have happened was by observers maintaining less stable criteria in the dissimilar condition than in the similar condition. An assumption of signal detection theory is that the observer's criterion is located at a fixed point on the x-axis. But there is every reason to suppose that just as the values of signal and noise vary according to Gaussian distributions, so also will each criterion have its own variance, as it is unlikely that observers will be able to make perfectly consistent judgements on all the trials.

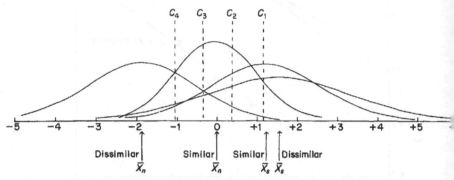

FIGURE 8.6 *The two signal distributions. the two noise distributions. and the four criteria for the letter recognition task rescaled on a common x-axis of S.D. units of the noise distribution for similar-non-targets*

Criterion variance was incorporated in Thurstone's Law of Comparative Judgement. Its effects have been extensively discussed by Ingleby (1968), and its consequences in signal detection tasks have been mentioned by Tanner (1961). To understand what a variable criterion will appear to do to the signal and noise distributions, look at formula (8.4(a)) which is the complete expression for the Law of Categorical Judgement (Torgerson, 1958).

$$X_c - X_i = z\sqrt{(\sigma_c^2 + \sigma_i^2 - 2r_{ci}\sigma_c\sigma_i)}. \qquad (8.4(a))$$

In this formula X_i is the mean of the distribution of sensory effect for stimulus S_i in a category rating task, or it might also be the mean of the signal or noise distribution in a conventional signal detection task. X_c is the mean of a criterion or category boundary, and σ_i and σ_c are the standard deviations of the distributions for the stimulus

200

and the criterion. The term z is the distance between the means, expressed as a standard normal deviate, and the term $2r_{ci}\sigma_c\sigma_i$ will be discussed presently. At the moment we will just note that r_{ci} is the correlation between S_i, the stimulus, and C, the criterion.

With slight re-arrangement the formula becomes

$$z = \frac{\overline{X}_c - \overline{X}_i}{\sqrt{(\sigma_c^2 + \sigma_i^2 - 2r_{ci}\sigma_c\sigma_i)}}. \qquad (8.4(b))$$

Formula (8.4(b)) is simply the distance between the means of the distributions for S_i and C, expressed in standard normal form.

A word now about the term $2r_{ci}\sigma_c\sigma_i$ which appears in the denominator of (8.4(b)). This is called a *covariance* term and its value will depend on the degree of correlation, r_{ci}, between the values of S_i and C. In Chapter 3 we recalled that the variance of a distribution of differences is equal to the sum of the variances of the two underlying distributions, but this is true only for distributions which are uncorrelated. When a non-zero correlation exists, the variance of the distribution of differences is equal to the sum of the variances of the two underlying distributions minus the co-variance. Thus the complete expression for the variance of a distribution of differences is

$$\sigma_{x_1-x_2}^2 = \sigma_{x_1}^2 + \sigma_{x_2}^2 - 2r_{x_1x_2}\sigma_{x_1}\sigma_{x_2}. \qquad (8.5)$$

If there is no correlation the co-variance term disappears and (8.5) becomes

$$\sigma_{x_1-x_2}^2 = \sigma_{x_1}^2 + \sigma_{x_2}^2. \qquad (8.6)$$

With no co-variance, the Law of Categorical Judgement, as expressed in (8.4(b)), becomes

$$z = \frac{\overline{X}_c - \overline{X}_i}{\sqrt{(\sigma_c^2 + \sigma_i^2)}}. \qquad (8.7)$$

In the following discussion of the Law of Categorical Judgement, it will be assumed that the co-variance term of (8.4) can be disregarded, and it will be with the version stated in (8.7), and simplifications of it, that we will be concerned.

A further simplification of (8.7) can be made if the criterion has no variance, and has a constant value, in the way signal detection theory assumes it does. The simplification is shown in (8.8).

$$z = \frac{C - \bar{X}_i}{\sigma_i}. \tag{8.8}$$

In this formula, C is the value of the fixed criterion.

If the stimulus distribution in question was the signal distribution in a conventional signal detection yes–no task. (8.8) would be the expression for $z(S\,|\,s)$, the distance of the criterion from the signal distribution mean, scaled in S.D. units of the signal distribution, that is

$$z(S\,|\,s) = \frac{C - \bar{X}_s}{\sigma_s}. \tag{8.9}$$

If the stimulus distribution in question was the noise distribution in a yes–no task, (8.8) would be the expression for $z(S\,|\,n)$, the distance of the criterion from the mean of the noise distribution, scaled in S.D. units of the noise distribution, that is

$$z(S\,|\,s) = \frac{C - \bar{X}_n}{\sigma_n} \tag{8.10}$$

What if we were wrong in assuming fixed criteria? How would this affect signal detection theory measures? Firstly, (8.9) and (8.10) would need to be modified to incorporate criterion variance in the manner of (8.7). These modified formulae are shown below

$$z(S\,|\,s) = \frac{\bar{X}_c - \bar{X}_s}{\sqrt{(\sigma_c^2 + \sigma_s^2)}} \tag{8.11}$$

$$z(S\,|\,n) = \frac{\bar{X}_c - \bar{X}_n}{\sqrt{(\sigma_c^2 + \sigma_n^2)}} \tag{8.12}$$

Unfortunately, in a conventional detection task with one signal distribution, one noise distribution, and a set of criteria, there is no way of disentangling criterion variance from signal and noise variance. In Chapter 4, when the unequal variance case was introduced, it was said that while there were four parameters of the underlying

202

distributions which might be changed (these being the two means and two variances for signal and noise) that two of these needed to be fixed, and the other two could only be measured relative to them. Conventionally then, it is the mean and standard deviation of the noise distribution which are held constant. Introducing a fifth variable, that of criterion variance, does not improve the situation in a normal detection task. If it is present, it will have the effect of artificially inflating the variance of signal and noise, from which it cannot be separated. The situation is illustrated in Figure 8.7. In Figure 8.7(a) the true state of affairs is shown for a task in which each of three criteria has a variance of σ_c^2, and the noise variance is set

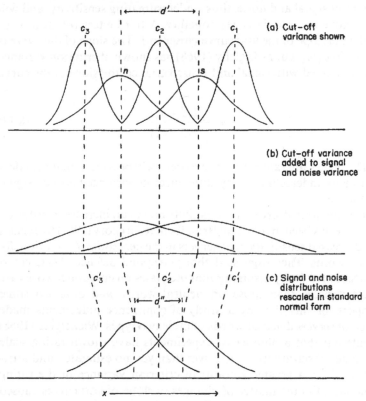

FIGURE 8.7   *Illustration of the effect of adding criterion variance to signal and noise variance*

203

equal to 1. If these criteria are incorrectly assumed to be constant, their variance is confounded with that of the signal and noise distributions, whose variances are both increased by a constant amount, σ_c^2, as is shown in Figure 8.7(b). By adding this extra source of variance to signal and noise, their distributions will no longer be in standard normal form, and when they are rescaled with the standard deviation of the noise distribution set equal to 1, the situation depicted in Figure 8.7(c) will occur. This shows that, from the point of view of an experimenter looking at data from a detection task where a signal distribution is scaled against its noise distribution, criterion variance will decrease the apparent distance between the means of signal and noise, thus underestimating sensitivity, and will also move the criteria closer together. A third effect will also be to make the slope of the ROC curve closer to 1. The slope of the curve is the ratio σ_n/σ_s, but as Ingleby (1968) has shown, if criterion variance is confounded with signal and noise variance, the slope of the curve will be

$$s = \frac{\sqrt{(\sigma_c^2 + \sigma_n^2)}}{\sqrt{(\sigma_c^2 + \sigma_s^2)}}. \tag{8.13}$$

The bigger the value of σ_c^2, the closer s will be to 1, so that the effect will be to underestimate any differences in variance between signal and noise.

The apparent decrease in sensitivity due to an increase in criterion variance has been invoked by Broadbent & Gregory (1967) to explain the decreased sensitivity of observers in the perception of emotionally toned words. They suggested that perceptual defence effects could be accounted for by assuming that the biases on emotional words are more variable than those on neutral words, and obtained some support for this view by a study of confidence judgements made when observers detected words of these two types. Wickelgren (1968) points out that a number of experiments have shown rating scale estimates of sensitivity to be lower than yes–no estimates, and attributes this to a greater amount of criterion variance in the rating scale task. On the matter of changes in slope of ROC curves caused by criterion variance, Lee (1963), using signal and noise distributions whose variances were known *a priori*, found that the slopes of ROC curves generated by a rating scale task were closer to unity than was predicted by the true σ_n/σ_s ratios.

The virtues of applying Thurstonian scaling techniques to yes–no detection tasks can also be seen by looking again at Figures 8.4 and 8.6 which show the distributions of signal and noise for the letter recognition task. In Figure 8.4 (conventional detection analysis) it is seen that the criteria in the dissimilar condition are closer together than those in the similar condition. When Thurstonian scaling is used (Figure 8.6) it is seen that assuming the criteria remain fixed in both conditions causes an apparent increase in the variances of signal and noise for the dissimilar condition by comparison with those of the similar condition. By looking at the σ-values in Figure 8.5, this variance increase is about 0·4 σ-units in both signal and noise, and it might be expected that this is an estimate of the difference in σ_c between the two conditions. It is still not possible to obtain absolute measures of σ_c any more than it is possible to obtain absolute measures of Δm, σ_s, or σ_n for any of the experimental conditions. In the example given, criterion variance in the dissimilar condition is measured relative to that in the similar condition, whose value of σ_c remains unknown. Thurstonian scaling techniques add extra versatility to signal detection analysis as they allow us to discard, or check, to a certain extent, the strict assumptions about the fixity of criteria and the unchanging nature of the noise distribution. It is worth remembering that these methods also make rather strict assumptions of their own. One of these is that at least the mean values of the criteria remain constant over the various experimental conditions whose signal and noise distributions are to be scaled on a common x-axis. If an experimental treatment changes the positions of the criteria themselves, the resulting scaling may be meaningless. A second assumption is that σ_c is constant for all criteria. But what would happen if different criteria had different variances? Ingleby (1968) has considered this possibility and says that the result would be to distort the shape of the ROC curve so that it would no longer be a straight line when plotted on a double-probability scale. On examining some of his own data, he found that the predicted distortions did occur, and that criteria associated with extreme confidence categories appear to be more variable than those for intermediate degrees of confidence. Fortunately the distortions were not large; if they were, signal detection theory would be greatly restricted in its usefulness.

FORCED-CHOICE TASKS AND THE LAW OF COMPARATIVE JUDGEMENT

Imagine, as was illustrated in Figure 8.1, that a number of stimuli can have their distributions of sensory effect scaled on a common x-axis. Rather than determining their relative sensitivities by the category rating task described earlier, we might just as well have paired off all the stimuli and conducted a series of 2AFC tasks. With the four stimuli illustrated in Figure 8.1 there would be twelve possible forced-choice tasks each yielding its own d'_{FC}. This is the sort of situation envisaged in Thurstone's (1927b) Law of Comparative Judgement which is embodied in the following formula

$$\bar{X}_i - \bar{X}_j = +z\sqrt{(\sigma_i^2 + \sigma_j^2 - 2r_{ij}\sigma_i\sigma_j)}. \tag{8.14(a)}$$

The Law states that the difference between the means of two distributions for stimuli S_i and S_j, when S_i is in the first interval of a 2AFC task and S_j in the second interval, is the product of a normal deviate, z, and the standard deviation of the distribution of differences, which is the portion of (8.14(a)) under the square root sign. The Law of Comparative Judgement closely resembles the Law of Categorical Judgement, as can be seen by comparing (8.4(a)) and (8.14(a)). The essential difference between the two Laws is that in categorical judgements it is the distance between the mean of a stimulus distribution and the mean of a criterion which is being determined; in the Law of Comparative Judgement the distance is between the means of two stimulus distributions.

Recalling the discussion of the 2AFC task in Chapter 3, it will be remembered that besides the distribution of differences for $\bar{X}_i - \bar{X}_j$, there will also be a distribution of differences for $\bar{X}_j - \bar{X}_i$ which will apply when S_j appears in the first interval of the 2AFC task, and S_i, in the second interval. For this case, the Law of Comparative Judgement will be as in (8.14(b)), differing from (8.14(a)) only in that $+z$ is changed to $-z$.

$$\bar{X}_j - \bar{X}_i = -z\sqrt{(\sigma_i^2 + \sigma_j^2 - 2r_{ij}\sigma_i\sigma_j)}. \tag{8.14(b)}$$

With some re-arrangement (8.14(a)) and (8.14(b)) become

$$+z = \frac{\bar{X}_i - \bar{X}_j}{\sqrt{(\sigma_i^2 + \sigma_j^2 - 2r_{ij}\sigma_i\sigma_j)}}. \tag{8.15(a)}$$

$$-z = \frac{\bar{X}_j - \bar{X}_i}{\sqrt{(\sigma_i^2 + \sigma_j^2 - 2r_{ij}\sigma_i\sigma_j)}}. \tag{8.15(b)}$$

These formulae are the means of the two distributions of differences expressed in standard normal form. From (8.15(a)) and (8.15(b)) we can find the distance between the means of the distributions of differences, and this distance will be the same as the measure d'_{FC} from signal detection theory.

$$d'_{FC} = \frac{2|\bar{X}_i - \bar{X}_j|}{\sqrt{(\sigma_i^2 + \sigma_j^2 - 2r_{ij}\sigma_i\sigma_j)}} \tag{8.16}$$

(The expression $|\bar{X}_i - \bar{X}_j|$ means the absolute value of $\bar{X}_i - \bar{X}_j$, i.e. it is always positive.)

If (8.16) does not look much like a familiar expression for d'_{FC} let us try a few simplifications.

First assume that there is no co-variance. In that case (8.16) becomes

$$d'_{FC} = \frac{2|\bar{X}_i - \bar{X}_j|}{\sqrt{(\sigma_i^2 + \sigma_j^2)}}. \tag{8.17}$$

Next assume that $\sigma_i = \sigma_j$ so that where σ_i appears in (8.17) we can write σ_j in its place. This gives us (8.18(a)).

$$d'_{FC} = \frac{2|\bar{X}_i - \bar{X}_j|}{\sqrt{(2\sigma_j^2)}}, \tag{8.18(a)}$$

and this can be rewritten as

$$d'_{FC} = \sqrt{2}\left(\frac{|\bar{X}_i - \bar{X}_j|}{\sigma_j}\right). \tag{8.18(b)}$$

Formula (8.18(b)) is interesting because if $\bar{X}_i$ was the mean of a signal distribution, $\bar{X}_s$, and $\bar{X}_j$ was the mean of a noise distribution, $\bar{X}_n$, and σ_j was the standard deviation of the noise distribution, the term $|\overline{X}_i - \overline{X}_j|/\sigma_j$ would be the expression for yes–no d' given in formula (3.1(b)). This being the case (8.18(b)) can be expressed as

$$d'_{FC} = d'\sqrt{2}. \qquad (8.18(c))$$

This shows that the familiar expression for forced-choice d' given in (8.18(c)), is a simplified version of Thurstone's Law of Comparative Judgement, and that the general expression for d'_{FC} is given in (8.16).

The simplifications involved in getting to (8.18(c)) from (8.16) are quite drastic, and involve assuming both that there is no correlation between the two stimulus intervals in the forced-choice task, and that the variances of the underlying stimulus distributions are equal. In reviewing some experiments which investigated the $d'\sqrt{2}$ relationship, Wickelgren (1968) says that it is truly amazing that the predicted relationship has held up so well.

Forced-choice experiments are popular because of the economy involved in collecting good data for sensitivity measures, and because of their comparative freedom from contamination by response bias. However, having obtained a forced-choice d' it is not normally possible to work back from it to the underlying distributions of signal and noise. Even by combining forced-choice methods with a rating scale, and thence by obtaining the ROC curve for the 2AFC experiment, we cannot deduce the signal distribution variance, because the ROC curve always has a slope of 1. This is because it is based on two distributions of differences whose variances will always be equal no matter how different were the variances of the original signal and noise distributions. However, by combining the Law of Comparative Judgement with signal detection theory's method of using a rating scale to determine the 2AFC ROC curve, it is possible to devise a situation where the means and standard deviations of the original signal and noise distributions might be inferred from 2AFC data, and where any co-variance might also be extracted.

Decomposition of 2AFC data

In this section a method will be outlined by which 2AFC data collected with confidence ratings can be decomposed into the means and variances of the underlying distributions of sensory effect, and the correlations between pairs of stimuli. Torgerson (1958) remarks that the Law of Comparative Judgement, as expressed in its complete form in formula (8.14), is unsolvable because there are always more unknowns than there are equations. Consequently, simplifying assumptions about the co-variances and the standard deviations of the type shown in (8.17) and (8.18) have needed to be used. Sjoberg (1967) has, however, shown that 2AFC data, collected with confidence judgements, can be used to extract the means of the underlying distributions, and the variances of the various distributions of differences. The method to be outlined here also relies on the use of confidence judgements, but in line with signal detection theory, uses these to generate 2AFC ROC curves the formulae for whose slopes and Δm values are used as a set of simultaneous equations from which means, variances and co-variances might be determined. It is similar to the technique by which Ingleby (1968) solves equations for category rating tasks, and will be called the *method of operating characteristics*.

We begin with n stimuli to be used in a series of 2AFC tasks. The first of these, S_1, is chosen as a reference point, and has a mean of 0 and an S.D. of 1. The means of all other stimuli are measured relative to the mean of S_1, and their standard deviations are ex-

TABLE 8.10 *Means and standard deviations for n stimuli relative to the mean and standard deviation for S_1*

Stimulus	Mean	Standard deviation
S_1	0	1
S_2	$\Delta m_2'$	σ_2
S_3	$\Delta m_3'$	σ_3
.		
.		
.		
S_j	$\Delta m_j'$	σ_j
.		
.		
S_k	$\Delta m_k'$	σ_k
.		
.		
S_n	$\Delta m_n'$	σ_n

pressed relative to the standard deviation of S_1. These means and standard deviations are shown in Table 8.10.

There will be $n(n-1)/2$ different ways in which the n stimuli can be paired off with each other. For each pairing of the stimuli S_i and S_j, there will be two differences, $S_i - S_j$ and $S_j - S_i$. For each such difference there will be a distribution of differences with its own mean and its own variance. These means and variances are shown in Table 8.11.

TABLE 8.11   *Means and variances for distributions of differences between n stimuli expressed in terms of the means and standard deviations of the underlying distributions*

Difference	Mean of distribution of differences	Variance of distribution of differences
$S_1 - S_2$	$-\Delta m'_2$	$1 + \sigma_2^2 - 2r_{12}\sigma_2$
$S_2 - S_1$	$+\Delta m'_2$	$1 + \sigma_2^2 - 2r_{12}\sigma_2$
$S_1 - S_3$	$-\Delta m'_3$	$1 + \sigma_3^2 - 2r_{13}\sigma_3$
$S_3 - S_1$	$+\Delta m'_3$	$1 + \sigma_3^2 - 2r_{13}\sigma_3$
.		
.		
.		
$S_2 - S_3$	$\Delta m'_2 - \Delta m'_3$	$\sigma_2^2 + \sigma_3^2 - 2r_{23}\sigma_2\sigma_3$
$S_3 - S_2$	$\Delta m'_3 - \Delta m'_2$	$\sigma_2^2 + \sigma_3^2 - 2r_{23}\sigma_2\sigma_3$
.		
.		
.		
$S_i - S_j$	$\Delta m'_i - \Delta m'_j$	$\sigma_i^2 + \sigma_j^2 - 2r_{ij}\sigma_i\sigma_j$
$S_j - S_i$	$\Delta m'_j - \Delta m'_i$	$\sigma_i^2 + \sigma_j^2 - 2r_{ij}\sigma_i\sigma_j$
$S_k - S_l$	$\Delta m'_k - \Delta m'_l$	$\sigma_k^2 + \sigma_l^2 - 2r_{kl}\sigma_k\sigma_l$
.		
.		
.		
$S_{n-1} - S_n$	$\Delta m'_{n-1} - \Delta m'_n$	$\sigma_{n-1}^2 + \sigma_n^2 - 2r_{(n-1),n}\sigma_{n-1}\sigma_n$
$S_n - S_{n-1}$	$\Delta m'_n - \Delta m'_{n-1}$	$\sigma_{n-1}^2 + \sigma_n^2 - 2r_{(n-1),n}\sigma_{n-1}\sigma_n$

The distributions of differences serve as the basis for the 2AFC ROC curves. In conventional signal detection analysis the observer is presented with drawings from the distributions for $S_i - S_j$ and $S_j - S_i$ in random order, and must rate his confidence as to which distribution the evidence came from. If the responses are called R_{i-j} and R_{j-i}, the ROC curve for the 2AFC task is a plot of $z(R_{i-j} | S_i - S_j)$

against $z(R_{i-j}|S_j-S_i)$. As both these distributions of differences have the same variance, the slope of the 2AFC ROC curve should be 1. There are other ROC curves that could be plotted. The plot of $z(R_{k-l}|S_k-S_l)$ against $z(R_{i-j}|S_i-S_j)$ will produce a 2AFC ROC whose slope need not necessarily be 1 because, as Table 8.11 shows, the variances of the distributions of differences for S_i-S_j and S_k-S_l may differ.

Pairing off each distribution of differences with each other one allows $n(n^3-2n^2+1)/2$ different ROC curves to be constructed. Each curve will have a slope, which will be the ratio of the standard deviations of the distributions of differences from which it was constructed. The equations for the squared values of the slopes, s^2, are shown in Table 8.12. As can be seen from the table, some of these slopes will be equal to 1, but others need not be. From each ROC curve a value of Δm can be obtained. As these ROC curves are for a forced-choice task, this sensitivity measure will be called Δm_{FC}.

Remembering that Δm is defined as the distance between the means of the two distributions on which the ROC curve is based, scaled in S.D. units of the distribution serving as the x-variable, a set of equations for Δm_{FC} can be written for all the 2AFC ROC curves. They are shown in Table 8.12.

For n stimuli there will be $n(n^3-2n^2+1)$ equations. Of these, half will be for the slopes of the ROC curves, and half for the Δm_{FC} values. There will be a total of $(n+4)(n-1)/2$ unknowns, $n-1$ each of $\Delta m'$ values and σs, and $n(n-1)/2$ of rs. So long as n is greater than 2, the system of simultaneous equations is over-determined, i.e. the number of equations exceeds the number of unknowns. In theory these equations are soluble for the unknowns, and it should therefore be theoretically possible to decompose 2AFC data collected with confidence judgements, into the means and standard deviations of the original distributions, and the correlations between the distributions.

The means by which such a set of simultaneous equations is solved will not be discussed here. The solution is only achievable by computer, and a number of programmes have been created for this purpose. They can be found by consulting the Collected Algorithms from the Communications of the Association for Computing Machinery. A more important question is: are the equations soluble in practice? There will always be error involved in estimating

211

TABLE 8.12   *Slopes and Δm values for 2AFC ROC curves expressed in terms of the means, standard deviations, and correlations for the underlying stimulus distributions*

Distributions of differences for 2AFC ROC curves		Slope² (s^2)	Δm_{FC}
x-variable	y-variable		
$S_1 - S_2$	$S_2 - S_1$	1	$\dfrac{-2\Delta m_2'}{\sqrt{(1 + \sigma_2^2 - 2r_{12}\sigma_2)}}$
$S_1 - S_2$	$S_1 - S_3$	$\dfrac{1 + \sigma_2^2 - 2r_{12}\sigma_2}{1 + \sigma_3^2 - 2r_{13}\sigma_3}$	$\dfrac{-\Delta m_2' - \Delta m_3'}{\sqrt{(1 + \sigma_2^2 - 2r_{12}\sigma_2)}}$
$S_1 - S_2$	$S_3 - S_1$	$\dfrac{1 + \sigma_2^2 - 2r_{12}\sigma_2}{1 + \sigma_3^2 - 2r_{13}\sigma_3}$	$\dfrac{-\Delta m_2' + \Delta m_3'}{\sqrt{(1 + \sigma_2^2 - 2r_{12}\sigma_2)}}$
$S_2 - S_3$	$S_3 - S_2$	1	$\dfrac{2(\Delta m_2' - \Delta m_3')}{\sqrt{(\sigma_2^2 + \sigma_3^2 - 2r_{13}\sigma_2\sigma_3)}}$
$S_2 - S_3$	$S_2 - S_4$	$\dfrac{\sigma_2^2 + \sigma_3^2 - 2r_{23}\sigma_2\sigma_3}{\sigma_2^2 + \sigma_4^2 - 2r_{24}\sigma_2\sigma_4}$	$\dfrac{\Delta m_4' - \Delta m_3'}{\sqrt{(\sigma_2^2 + \sigma_3^2 - 2r_{23}\sigma_2\sigma_3)}}$
$S_i - S_j$	$S_j - S_i$	1	$\dfrac{2(\Delta m_i' - \Delta m_j')}{\sqrt{(\sigma_i^2 + \sigma_j^2 - 2r_{ij}\sigma_i\sigma_j)}}$
$S_i - S_j$	$S_k - S_l$	$\dfrac{\sigma_i^2 + \sigma_j^2 - 2r_{ij}\sigma_i\sigma_j}{\sigma_k^2 + \sigma_l^2 - 2r_{kl}\sigma_k\sigma_l}$	$\dfrac{\Delta m_i' - \Delta m_j' - \Delta m_k' + \Delta m_l'}{\sqrt{(\sigma_i^2 + \sigma_j^2 - 2r_{ij}\sigma_i\sigma_j)}}$
$S_n - S_{n-1}$	$S_{n-1} - S_n$	1	$\dfrac{2(\Delta m_n' - \Delta m_{n-1}')}{\sqrt{(\sigma_n^2 + \sigma_{n-1}^2 - 2r_{n,(n-1)}\sigma_n\sigma_{n-1})}}$

the slopes and Δm_{FC} values from ROC curves, and it may be too great to allow the extraction of the means, standard deviations, and correlations from the 2AFC task.

Sjoberg (1967) has applied his version to the solution of the Law of Comparative Judgement to the scaling of nine immoral actions. From 2AFC rating data he obtained estimates of the means of the underlying distributions which correlated well with estimates obtained by the method of successive intervals, described in the previous section of this chapter. It would seem hopeful then that at least the formulae for Δm_{FC} in Table 8.12 could be used to work back to the original $\Delta m'$ values. In addition there are also the equations for the slopes of the ROC curves which may be able to be used in estimating the standard deviations of the original distributions, and the correlations between distributions. Ingleby (1968) certainly had some success in using the slopes of yes–no ROC curves to find variances of signal and noise distributions, but of course the 2AFC slope equations are more complicated than those for the yes–no task. It may be necessary to assume that the co-variance terms in the equations can be omitted, in which case only the σs would need to be solved for.

One final remark: although this method of dealing with 2AFC data does not necessitate the strict assumptions normally adopted by signal detection theory, some assumptions have still been made. As a rating scale has been introduced, there are the same possibilities for criterion variance, and co-variance between criteria and the distributions of differences which are detailed in the Law of Comparative Judgement. Thus to obtain a complete solution to the Law of Comparative Judgement, strict assumptions need to be made about the Law of Categorical Judgement. There is no way out of this problem, and if it appears that one set of implausible assumptions is being exchanged for another, it might be consoling to remember that even the least plausible and strictest assumptions that can be made about the 2AFC task have produced reasonable results.

Problems

1. The confusion matrix below is for a category rating task with three stimuli and four response categories. Find $\Delta m'$ and σ for each stimulus using the distribution for Stimulus 1 as the reference distribution.

		Response category			
		1	2	3	4
Stimulus event	1	69	24	5	2
	2	40	20	9	31
	3	37	13	7	43

2. In a yes–no task a signal and a noise distribution have equal variances, and $d' = 1$. If the variance of the criterion is equal to that of the noise distribution, what will the observed value of d' be if criterion variance is confounded with signal and noise variance?

3. If the standard deviation of a signal distribution is $2·0$, the standard deviation of a noise distribution is $1·0$, and the criterion standard deviation is $1·5$, what will the true and observed slopes of the ROC curve be?

4. Three stimuli, S_1, S_2, and S_3 have means of 0, 1, and 2, and standard deviations of 1, 2, and 3 respectively. Assuming no co-variance between any pairs of distributions, find the slopes and Δm_{FC} values for all possible ROC curves based on the six distributions of differences.

BIBLIOGRAPHY

ALLEN, L. R., & GARTON, R. F. (1968), 'The influence of word-knowledge on the word-frequency effect in recognition memory', *Psychonomic Science*, **10**, 401-2.

ALLEN, L. R., & GARTON, R. F. (1969), 'Detection and criterion change associated with different test contexts in recognition memory'. *Perception & Psychophysics*. **6**, 1-4.

BANKS, W. P. (1970), 'Signal detection theory and human memory', *Psychological Bulletin*, **74**, 81-99.

BLACKWELL, H. R. (1953), 'Psychophysical thresholds: experimental studies on methods of measurement', *Bull. Eng. Res. Inst. Univ. Mich.*, No. 36, Ann Arbor: University of Michigan Press.

BONEAU, C. A., & COLE, J. L. (1967), 'Decision theory, the pigeon, and the psychophysical function', *Psychological Review*, **74**, 123-35.

BROADBENT, D. E. (1967), 'Word-frequency effect and response bias', *Psychological Review*, **74**, 1-15.

BROADBENT, D. E., & GREGORY, M. (1963a), 'Vigilance considered as a statistical decision', *British Journal of Psychology*, **54**, 309-23.

BROADBENT, D. E., & GREGORY, M. (1963b), 'Division of attention and the decision theory of signal detection,' *Proc. Roy. Soc. B*, **158**, 222-31.

BROADBENT, D. E., & GREGORY, M. (1965), 'Effects of noise and of signal rate on vigilance analysed by means of decision theory', *Human Factors*, **7**, 155-62.

BROADBENT, D. E., & GREGORY, M. (1967), 'Perception of emotionally toned words', *Nature*; **215**, 581-4.

BROWN, J. (1964), 'Two tests of all-or-none learning and retention', *Quarterly Journal of Experimental Psychology*, **16**, 123-33.

BROWN, J. (1965), 'Multiple response evaluation of discrimination', *British Journal of Mathematical & Statistical Psychology*, **18**, 125-37.

BUSHKE, H. (1963), 'Relative attention in immediate memory determined by the missing scan method', *Nature*, **200**, 1129-30.

CONKLIN, E. S. (1923), 'The scale of values method for studies in genetic psychology', *Univ. Ore. Pub.*, **2**, No. 1. (Cited by Guilford, 1936).

COOMBS, C. H., DAWES, R. M., & TVERSKY, A. (1970), *Mathematical psychology*, New Jersey: Prentice-Hall.

CORSO, J. F. (1963), 'A theoretico-historical review of the threshold concept', *Psychological Bulletin*, **60**, 356-70.

DANDELIKER, J., & DORFMAN, D. D. (1969), 'Receiver-operating characteristic curves for taboo and neutral words', *Psychonomic Science*, **17**, 201-2.

215

P

DEMBER, W. N. (1964), *The psychology of perception*, New York: Holt, Rinehart & Winston.

DONALDSON, W. & MURDOCK, B. B. JR. (1968), 'Criterion changes in continuous recognition memory', *Journal of experimental Psychology*. **76**, 325–30.

DORFMAN, D. D. (1967), 'Recognition of taboo words as a function of *a priori* probability', *J. Pers. Soc. Psychol.*. **7**, 1–10.

DORFMAN, D. D., & ALF, E. JR. (1968), 'Maximum likelihood estimation of parameters of signal detection theory—a direct solution', *Psychometrika*, **33**, 117–24.

DORFMAN, D. D., & ALF, E. JR. (1969), Maximum-likelihood estimation of parameters of signal detection theory and determination of confidence intervals-rating-method data, *J. Math. Psychol.*, **6**, 487–96.

EGAN, J. P. (1958), 'Recognition memory and the operating characteristic', *Indiana University: Hearing and Communication Lab. Tech. Rep.*, No. AFCRC-TR-57-50.

EGAN, J. P., & CLARKE, F. R. (1966), 'Psychophysics and signal detection', In: Sidowski, J. B. (Ed.), *Experimental methods and instrumentation in psychology*, New York: McGraw-Hill, 211–46.

EGAN, J. P., SCHULMAN, I., & GREENBERG, G. Z. (1959), 'Operating characteristics determined by binary decisions and by ratings', *J. Acoust. Soc. Amer.*. **31**, 768–73.

ELLIOTT, P. B. (1964), 'Tables of *d'* ', in Swets, J. A. (Ed.), *Signal detection and recognition by human observers*, New York: Wiley. 651–84.

FITZHUGH, R. (1957), 'The statistical detection of threshold signals in the retina', *J. gen. Physiol.*, **40**, 925–48.

GALANTER, E., & MESSICK, S. (1961), 'The relation between category and magnitude scales of loudness', *Psychological Review*. **68**, 363–72.

GARNER, W. R. (1962), *Uncertainty and structure as psychological concepts*, New York: Wiley.

GOLDIAMOND, I., & HAWKINS, W. F. (1958), 'Vexierversuch: The log relationship between word-frequency and recognition obtained in the absence of stimulus words', *Journal of Experimental Psychology*. **56**, 457–63.

GREEN, D. M., & BIRDSALL, T. G. (1964), 'The effect of vocabulary size on articulation score', in Swets, J. A. (Ed.), *Signal detection and recognition by human observers*. New York:Wiley. 609–19

GREEN, D. M., & SWETS, J. A. (1966), *Signal detection theory and psychophysics*. New York: Wiley.

GUILFORD, J. P. (1936), *Psychometric methods*. New York: McGraw-Hill.

HACKHILL, M. H. (1963). 'Signal Detection in the rat', *Science* **139**, 758–9.

HARDY, G. R., & LEGGE, D. (1968), 'Cross-modal induction of changes in sensory thresholds', *Quarterly Journal of Experimental Psychology*. **20**, 20–9.

HULL, C. L. (1943), *Principles of behaviour*. New York: Appleton-Century-Crofts.

INGLEBY, J. D. (1968), *Decision-making processes in human perception and memory*. Unpublished Ph.D. thesis: University of Cambridge.

JOHN, I. D. (1967), 'A statistical decision theory of simple reaction time', *Australian Journal of Psychology*, **19**. 27–34.

LEE, W. (1963), 'Choosing among confusably distributed stimuli with specified likelihood ratios', *Perceptual Motor Skills*, **16**, 445–67.

LEE, W. (1969), 'Relationships between Thurstone category scaling and signal detection theory', *Psychological Bulletin*, **71**, 101–7.

LOCKHART, R. S., & MURDOCK, B. B. JR. (1970), 'Memory and the theory of signal detection', *Psychological Bulletin*, **74**, 100–9.

Luce, R. D. (1959), *Individual choice behaviour*, New York: Wiley.
Luce, R. D. (1963), 'Detection and recognition of human observers', in Luce, R. D., Bush, R. R., & Galanter, E. (1963) *Handbook of mathematical psychology*, Vol. I., New York: Wiley.
Mackworth, J. F., & Taylor, M. M. (1963), 'The d' measure of detectability in vigilance-like situations', *Canadian Journal of Psychology*, 17, 302-25.
Markowitz, J., & Swets, J. A. (1967), 'Factors affecting the slope of empirical ROC curves: comparisons of binary and rating responses', *Perception & Psychophysics*, 2, 91-100.
Miller, G. A., Heise, G. A., & Lichten, W. (1951), 'The intelligibility of speech as a function of the context of the test materials', *Journal of Experimental Psychology*, 41, 329-35.
Moray, N., & O'Brien, T. (1967), 'Signal detection theory applied to selective listening, *J. acoust. Soc. Amer.*, 42, 765-72.
Murdock, B. B. Jr. (1963), 'An analysis of the recognition process', in Cofer, C. N., & Musgrave, B. S. (Eds.) *Verbal behaviour and learning: problems and processes*, New York: McGraw-Hill. 10-22.
Murdock, B. B. Jr. (1965), 'Signal detection theory and short-term memory', *Journal of Experimental Psychology*, 70, 443-7.
Murdock, B. B. Jr. (1968), 'Serial order effects in short-term memory', *Journal of Experimental Psychology Monograph Supplement*, 76, No. 4, Part 2.
McNicol, D. (1971), 'The confusion of order in short-term memory', *Australian Journal of Psychology*. (In press)
McNicol, D., & Willson, R. J. (1971), 'The application of signal detection theory to letter recognition', *Australian Journal of Psychology*. (In press).
Nevin, J. A. (1964), 'A method for the determination of psychophysical functions in the rat', *J. Exp. Anal. Behaviour*, 7, 169.
Nevin, J. A. (1965), 'Decision theory in studies of discrimination in animals'. *Science*, 150, 1057.
Norman, D. A. (1964), 'A comparison of data obtained with different false-alarm rates', *Psychological Review*, 71, 243-6.
Norman, D. A., & Wickelgren, W. A. (1965), 'Short-term recognition memory for single digits and pairs of digits', *Journal of Experimental Psychology*, 70, 479-89.
Ogilvie, J. C., & Creelman, C. D. (1968), 'Maximum-likelihood estimation of Receiver Operating Characteristic curve parameters', *J. math. Psychol.*, 5, 377-91.
Pinneo, L. R. (1966), 'On noise in the nervous system', *Psychological Review*, 73, 242-7.
Pollack, I., & Decker, L. R. (1958), 'Confidence ratings, message reception and the receiver operating characteristic', *J. acoust. Soc. Amer.*, 30, 286-92.
Pollack, I., & Norman, D. A. (1964), 'A non-parametric analysis of recognition experiments', *Psychonomic Science*, 1, 125-6.
Pollack, I., Norman, D. A., & Galanter, E. (1964), 'An efficient non-parametric analysis of recognition memory', *Psychonomic Science*, 1, 327-8.
Price, R. H. (1966), 'Signal detection methods in personality and perception', *Psychological Bulletin*, 66, 55-62.
Rilling, M., & McDiarmid, C. (1965), 'Signal detection in fixed-ratio schedules', *Science*, 148, 526-7.
Runquist, W. N. (1966), 'Verbal behaviour', in Sidowski, J. B. (Ed.), *Experimental methods and instrumentation in psychology*, New York: McGraw-Hill, 478-540.

217

SAFFIR, M. A. (1937), 'A comparative study of scales constructed by three psychophysical methods', *Psychometrika*, 2, 179–98.

SCHÖNEMANN, P. H., & TUCKER, L. R. (1967), 'A maximum likelihood solution for the method of successive intervals allowing for unequal stimulus dispersions', *Psychometrika*, 32, 403–17.

SCHULMAN, A. I., & MITCHELL, R. R. (1966), 'Operating characteristics from yes–no and forced-choice procedures', *J. acoust. Soc. Amer.*, 40, 473–7.

SJÖBERG, L. (1967), 'Successive intervals scaling of paired comparisons', *Psychometrika*, 32, 297–308.

STOWE, A. N., HARRIS, W. P., & HAMPTON, D. B. (1963), 'Signal and context components of word recognition behaviour', *J. acoust. Soc. Amer.*, 35, 639–44.

SUBOSKI, M. D. (1967), 'The analysis of classical discrimination conditioning experiments', *Psychological Bulletin*, 68, 235–42.

SWETS, J. A. (1959), 'Indices of signal detectability obtained with various psychophysical procedures', *J. acoust. Soc. Amer.*, 31, 511–13.

SWETS, J. A. (1961), 'Is there a sensory threshold?', *Science*, 134, 168–77.

SWETS, J. A. (Ed.) (1964), *Signal detection and recognition by human observers*, New York: Wiley.

SWETS, J. A., TANNER, W. P. JR., & BIRDSALL, T. G. (1961), 'Decision processes in perception', *Psychological Review*, 68, 301–40.

SYMONDS, P. M. (1924), 'On the loss in reliability in ratings due to coarseness of the scale', *Journal of Experimental Psychology*, 7, 456–61.

TANNER, W. P. JR. (1961), 'Physiological implications of psychophysical data', *Science*, 89, 752–65.

TEGHTSOONIAN, R. (1965), 'The influence of number of alternatives on learning and performance in a recognition task', *Canadian Journal of Psychology*, 19, 31–41.

THORNDIKE, E. L., & LORGE, I. (1944), *The teacher's word book of 30,000 words*, New York: Teachers College, Columbia University, Bureau of Publications.

THURSTONE, L. L. (1927a), 'Psychophysical analysis', *American Journal of Psychology*, 38, 368–89.

THURSTONE, L. L. (1927b), 'A Law of Comparative Judgement', *Psychological Review*, 34, 273–86.

TIPPETT, L. H. C. (1925), 'On the extreme individuals and the range of samples taken from a normal population', *Biometrika*, 17, 364–87.

TORGERSON, W. S. (1958), *Theory and methods of scaling*, New York: Wiley.

TREISMAN, M., & WATTS, T. R. (1966), 'Relation between signal detectability theory and the traditional procedures for measuring sensory thresholds: estimating d' from results given by the method of constant stimuli', *Psychological Bulletin*, 66, 438–54.

WATSON, C. S., RILLING, M. E., & BOURBON, W. T. (1964), 'Receiver-operating characteristics determined by a mechanical analog to the rating scale', *J. acoust. Soc. Amer.*, 36, 283–8.

WELFORD, A. T. (1968), *Fundamentals of skill*, London: Methuen.

WELFORD, A. T. (1958), *Ageing and human skill*, Oxford: Oxford University Press.

WICKELGREN, W. A. (1968), 'Unidimensional strength theory and component analysis of noise in absolute and comparative judgements', *J. math. Psychol.*, 5, 102–22.

WICKELGREN, W. A., & NORMAN, D. A. (1966), 'Strength models and serial position in short-term recognition memory', *J. math. Psychol.*, 3, 316–47.

Appendix 1

ANSWERS TO PROBLEMS

CHAPTER 1

1. 0·67.
2. Respond S when $x \geqslant 4$; respond N when $x \leqslant 3$.
3. $\beta = 2$. Respond S when $x \geqslant 5$; respond N when $x < 5$.
4. $P(S|s) = 0·67, P(S|n) = 0·33$.
5. $P(N|s) = 0·11; \beta = \frac{1}{2}$ or $\frac{2}{3}$.
6. Respond S when $x \geqslant 2$; respond N when $x = 1$. $\beta = \frac{1}{2}$.
7. $l(x) = 2·50, 1·75, 1·60, 1·50$.
8. $P(x|S) = 0·15$.
9. $P(N|s) = 0·3, P(S|n) = 0·6$.
10. $l(x) = 1, \frac{1}{2}, \frac{2}{3}, \frac{1}{2}, \frac{4}{3}, 1, 2, \infty, \infty$.
(a) Respond S if $x \leqslant 63$ or if $x \geqslant 67$. Respond N if $64 \leqslant x \leqslant 66$.
(b) Respond S if $x \geqslant 65$; respond N if $x < 65$.

CHAPTER 2

1.
Criterion:	Strict	Medium	Lax
$P(S|s)$ =	0·50	0·70	0·90
$P(S|n)$ =	0·14	0·30	0·70

(a) 0·74.
2. (a) Observer 1:
$P(S|s) = 0·31, 0·50, 0·69, 0·77, 0·93, 0·98, 1·00.$
$P(S|n) = 0·02, 0·07, 0·16, 0·23, 0·50, 0·69, 1·00.$
Observer 2:
$P(S|s) = 0·23, 0·44, 0·50, 0·60, 0·69, 0·91, 1·00.$
$P(S|n) = 0·11, 0·26, 0·31, 0·40, 0·50, 0·80, 1·00.$
(b) Observer 1 = 0·85; observer 2 = 0·63.

(c) 0·81, 0·83, 0·85, 0·83, 0·81.

3. | x | 1 | 2 | 3 | 4 | 5 | 6 | 7 | 8 | 9 | 10 |
|---|---|---|---|---|---|---|---|---|---|---|
| $P(x \mid s)$ | 0·00 | 0·06 | 0·07 | 0·13 | 0·14 | 0·20 | 0·14 | 0·13 | 0·07 | 0·06 |
| $P(x \mid n)$ | 0·06 | 0·07 | 0·13 | 0·14 | 0·20 | 0·14 | 0·13 | 0·07 | 0·06 | 0·00 |

4. Observer:

1	2	3	4	5	6	7	8	9	10	11	12
B	B	B	B	B	B	B	U	B	W	B	B

4. 0·72.

CHAPTER 3

1. (a) 0·52; (b) 1·01; (c) 0·72.
2. (a) 1·31; (b) 1·72; (c) 3·39.
3. (a) +0·44; (b) +0·89.
4. (a) 0·84; (b) 0·11; (c) 0·80; (d) 0·69; (e) 0·86.
5. b and c.
6. (a) 1·56; (b) 1·36; (c) 0·61.
7. (a) −2·10; (b) +1·55; (c) −4·42.
8. (a) 1·68 or 0; (b) 0·80 or 0·20.
9. Hypothesis b.
10. $z(S \mid s) = +0·15, −0·41, −0·71, −1·48$.
    $z(S \mid n) = +1·65, +1·08, +0·81, +0·05$.
    $d' = 1·5$.
11. (a) 0·67; (b) 0·95.
12. $d'_{FC} = 1·41, d' = 1$.

CHAPTER 4

1. (a) −0·71; (b) −5·0; (c) +2·25.
2. (a) $\Delta m = 1·2, d'_e = 1·8$; (b) $\Delta m = 2·0, d'_e = 1·2$
   (c) $\Delta m = 1·6, d'_e = 1·3$.
3. (a) 0·69; (b) 0·16.
4. (a) 1·5; (b) 0·5; (c) 2·3; (d) 0·7.
5. (a) 0·80; (b) 0·09; (c) 0·25.
6. No.
7. (a) 1·02; (b) 2·39; (c) 1·70; (d) 1·39; (e) 1·42.
8. (a) 1; (b) 1·3; (c) 0·5; (d) 3·9, 3·5, 2·1, 1·1; (e) $z(S \mid s) = +2·58$,
   $z(S \mid n) = +2·29$; (f) $P(S \mid s) = 0·98, P(S \mid n) = 0·50$.

CHAPTER 5

1. (a) Observer 1 = 0·8, observer 2 = 1·2, observer 3 = 0·8.
   (b) Observer 1 = 1·0, observer 2 = 1·0, observer 3 = 0·6.
   (c) Observer 1 = 0·75, observer 2 = 0·73, observer 3 = 0·66.
2. $P(A)$: Obs. 1 = 0·88, obs. 2 = 0·88, obs. 3 = 0·71, obs. 4 = 0·73.
   $P(B)$: Obs. 1 = 3·11, obs. 2 = 1·70, obs. 3 = 3·17, obs. 4 = 1·64.

CHAPTER 6

1. (a) 0·76; (b) 0·15; (c) 0·20; (d) 0·72.
2. (a) $P(S|s) = 0·60, P(S|n) = 0·04$;
   (b) $P(N|s) = 0·20, P(N|n) = 0·50$;
   (c) $\alpha = 3·06, v = 1·31$;
   (d) $P(S|s) = 0·97, P(S|n) = 0·76$;
   (e) $\bar{X}_s - \bar{X}_n = 0·37, x = +0·37$;
   (f) $+1·48$.
3. (a) 80·5; (b) 122·5; (c) $v_a = 9·6, v_b = 6·02$.
4. (a) $\alpha_A^* = 14·2, \alpha_B^* = 256·0, \alpha_C^* = 1·7, \alpha_D^* = 12·8$.
   (b) $v_A = 10, v_C = 80, v_D = 10$.

CHAPTER 7

1. 0·80, 0·50, 0·35, 0·25, 0·20.
2. $\bar{L} = 0·60; d' = 0·27, 1·33, 1·83, 2·82, 3·32, 4·38$.
3. $p = 0·36, 0·51, 0·42; d'_{FC} = 1·41$.
4. $P(c) = \dfrac{\alpha}{\alpha + m - 1}$
5. No, although choice theory does better than high threshold theory, which is more noticeable if $\log \alpha$ values (approximately proportional to d') are taken.
   $p = 0·52, 0·41, 0·29, 0·21, 0·13, \ 0·04$.
   $\alpha = 3·17, 3·82, 4·29, 5·27, 5·90, 10·63$.

CHAPTER 8

1.
Stimulus:	1	2	3
$\Delta m'$:	0	+1	+1·5
σ:	1	2	3

221

2. 0.71.

3. $0.50, 0.87$.

4.   Distributions of differences for 2AFC ROC curves

x-variable	y-variable	Slope	Δm_{FC}
$S_1 - S_2$	$S_2 - S_1$	1	-0.90
$S_1 - S_2$	$S_1 - S_3$	0.71	$+0.45$
$S_1 - S_2$	$S_3 - S_1$	0.71	-1.34
$S_1 - S_2$	$S_2 - S_3$	0.62	0
$S_1 - S_2$	$S_3 - S_2$	0.62	-0.90
$S_2 - S_1$	$S_1 - S_3$	0.71	$+1.34$
$S_2 - S_1$	$S_3 - S_1$	0.71	-0.45
$S_2 - S_1$	$S_2 - S_3$	0.62	$+0.45$
$S_2 - S_1$	$S_3 - S_2$	0.62	0
$S_1 - S_3$	$S_3 - S_1$	1	-1.27
$S_1 - S_3$	$S_2 - S_3$	0.88	-0.32
$S_1 - S_3$	$S_3 - S_2$	0.88	-0.95
$S_3 - S_1$	$S_2 - S_3$	0.88	$+0.95$
$S_3 - S_1$	$S_3 - S_2$	0.88	$+0.32$
$S_2 - S_3$	$S_3 - S_2$	1	$+0.56$

Appendix 2

LOGARITHMS

The main virtue of logarithms is that they enable multiplication to be done by addition, and division to be done by subtraction, thus simplifying calculations. Two types of logarithms will be encountered in this book, *common logarithms* (or logarithms to the base 10) and *natural logarithms* (or logarithms to the base e).

COMMON LOGARITHMS

The definition of a common logarithm is as follows:

The common logarithm of a number, x, is the power to which 10 must be raised to produce that number x.

This definition can be expressed symbollically thus:

$$x = 10^{\log x}$$

Hence the expression $\log 2 = 0.3010$ is equivalent to the expression $10^{0.3010} = 2$.

When working with logarithms, addition can be substituted for multiplication. Thus

$$\log(5 \times 3) = \log 5 + \log 3.$$

and

$$\log 1760 = \log(1.76 \times 10^3) = \log 1.76 + \log 10^3$$
$$= \log 1.76 + 3 \log 10 = \log 1.76 + 3.$$

Similarly subtraction can be substituted for division. Thus

$$\log\left(\tfrac{5}{3}\right) = \log 5 - \log 3$$

and

$$\log 0{\cdot}00176 = \log(1{\cdot}76/10^3) = \log 1{\cdot}76 - \log 10^3$$
$$= \log 1{\cdot}76 - 3.$$

In this last case the logarithm of the number is usually written with the negative portion preceding the decimal place and the positive portion following it,

$$\log 0{\cdot}00176 = \bar{3}{\cdot}2455$$

However it is sometimes convenient for the purposes of calculation to express this as a *negative logarithm* by subtracting the positive part of the logarithm from the negative part.

$$\log 0{\cdot}00176 = 0{\cdot}2455 - 3 = -2{\cdot}7545$$

NATURAL LOGARITHMS

Natural logarithms use the base e instead of the base 10, where e is the mathematical constant approximately equal to $2{\cdot}718$. In this book we will use the convention of writing the common logarithm of x as $\log x$ and the natural logarithm of x as $\ln x$.

Just as for $\log x$, $\ln x$ can be defined symbolically as follows:

$$x = e^{\ln x}$$

As $\ln e = 1$ it follows that $x \ln e = x$ or

$$x = \ln e^x$$

Division and multiplication can be carried out in the same way by natural logarithms as by common logarithms.

224

Appendix 3

INTEGRATION OF THE EXPRESSION FOR THE LOGISTIC CURVE

To show that

$$\int_x^m \frac{e^r}{(1+e^x)^2} = \frac{e^x}{1+e^x}$$

the easiest plan is to differentiate the integral. Thus we wish to show that;

$$\frac{d}{dx}\left[\frac{e^x}{1+e^x}\right] = \frac{e^x}{(1+e^x)^2}$$

If we let $e^x = u$ and $(1+e^x) = v$, the differential is of the standard form

$$\frac{d}{dx}\left[\frac{u}{v}\right] = \frac{1}{v^2}\left[v\frac{du}{dx} - u\frac{dv}{dx}\right]$$

Hence

$$\frac{d}{dx}\left[\frac{e^x}{1+e^x}\right] = \frac{1}{(1+e^x)^2}\left[(1+e^x)\frac{d}{dx}e^x - e^x\frac{d}{dx}(1+e^x)\right].$$

As $\dfrac{d}{dx}e^x = e^x$ and $\dfrac{d}{dx}1 = 0$, the expression becomes

$$\frac{1}{(1+e^x)^2}\left[(1+e^x)e^x - e^{2x}\right] = \frac{e^x}{(1+e^x)^2}$$

This page intentionally left blank

Appendix 4

TABLES

I. TABLE OF $P(\bar{A})$

(The approximate area under an ROC curve estimated from a single pair of $P(S\,|\,s)$ and $P(S\,|\,n)$ values.)

Instructions for use

Take whichever is the larger of $P(S\,|\,s)$ and $P(S\,|\,n)$ and locate its value in the rows of the table. Then find the column entry corresponding to the smaller of the two values.

If $P(S\,|\,s)$ is greater than $P(S\,|\,n)$ the number in the cell designated by the row and column values is $P(\bar{A})$. If $P(S\,|\,n)$ is greater than $P(S\,|\,s)$, $P(\bar{A})$ is 1 minus the cell entry.

Note that decimal points have been omitted before the first digit of each number.

P_s, the smaller of the two probabilities increases by steps of 0·01. P_l, the larger of the two probabilities goes up in steps of 0·05 with the exception of the range 0·01 to 0·05 and the value 0·99. If the obtained value of P_l falls between two rows in the table it can still be obtained with reasonable accuracy by extrapolation.

For experimenters who require more accurate estimates of $P(\bar{A})$ than this table provides, the following formula may be used.

$$P(\bar{A}) = \tfrac{1}{2}\left(\frac{x^3 + x^2(y-2) - x(y+3)(y-1) - y(y-1)^2}{2x(1-y)}\right)$$

where $x = P(S\,|\,s)$ and $y = P(S\,|\,n)$.

(These tables were prepared by a computer programme written by Su Williams from the Department of Psychology, University of Adelaide.)

227

P_t	01	02	03	04	05	06	07	08	09	P_s 10	11	12	13	14	15	16	17	18	19	2
01	50																			
02	63	50																		
03	67	59	50																	
04	70	63	57	50																
05	71	66	61	55	50															
10	75	72	69	67	64	61	58	56	53	50										
15	77	75	73	71	69	67	66	64	62	60	58	56	54	52	50					
20	79	77	76	74	73	71	70	68	67	65	64	62	61	59	58	56	55	53	52	5
25	80	79	78	77	75	74	73	72	70	69	68	67	65	64	63	62	60	59	58	5
30	82	81	80	78	77	76	75	74	73	72	71	70	69	68	67	66	65	64	63	6
35	83	82	81	80	79	78	78	77	76	75	74	73	72	71	70	69	68	67	66	6
40	84	83	83	82	81	80	80	79	78	77	76	76	75	74	73	72	71	71	70	6
45	86	85	84	84	83	82	81	81	80	79	78	78	77	76	76	75	74	73	73	7
50	87	86	86	85	84	84	83	82	82	81	81	80	79	79	78	77	76	76	75	7
55	88	88	87	87	86	85	85	84	84	83	82	82	81	81	80	79	79	78	78	7
60	90	89	88	88	87	87	86	86	85	85	84	84	83	83	82	81	81	80	80	7
65	91	90	90	89	89	88	88	87	87	86	86	85	85	84	84	83	83	82	82	8
70	92	92	91	91	90	90	89	89	89	88	88	87	87	86	86	85	85	84	84	8
75	93	93	93	92	92	91	91	91	90	90	89	89	89	88	88	87	87	86	86	8
80	95	94	94	94	93	93	92	92	92	91	91	91	90	90	89	89	89	88	88	8
85	96	96	95	95	95	94	94	94	93	93	93	92	92	92	91	91	91	90	90	
90	97	97	97	96	96	96	95	95	95	94	94	94	94	93	93	93	92	92	92	9
95	99	98	98	98	97	97	97	97	96	96	96	95	95	95	95	94	94	94	94	
99	100	99	99	99	99	98	98	98	98	97	97	97	96	96	96	96	95	95	95	9

P_t	21	22	23	24	25	26	27	28	29	P_s 30	31	32	33	34	35	36	37	38	39	4
25	55	54	53	51	50															
30	60	59	58	57	56	55	54	52	51	50										
35	64	64	63	62	61	59	59	57	56	55	54	53	52	51	50					
40	68	67	66	65	64	64	63	62	61	60	59	58	57	56	55	54	53	52	51	
45	71	70	69	69	68	67	66	65	65	64	63	62	61	60	59	59	58	57	56	
50	74	73	72	72	71	70	69	69	68	67	66	66	65	64	63	63	62	61	60	
55	76	76	75	74	74	73	72	72	71	70	70	69	68	68	67	66	65	65	64	
60	79	78	77	77	76	76	75	74	74	73	73	72	71	71	70	69	69	68	67	
65	81	80	80	79	79	78	78	77	77	76	75	75	74	74	73	73	72	71	71	
70	83	83	82	82	81	81	80	80	79	79	78	78	77	77	76	75	75	74	74	
75	85	85	84	84	83	83	82	82	82	81	81	80	80	79	79	78	78	77	77	
80	87	87	86	86	86	85	85	84	84	84	83	83	82	82	81	81	81	80	80	
85	89	89	88	88	88	87	87	87	86	86	85	85	85	84	84	84	83	83	82	
90	91	91	90	90	90	89	89	89	88	88	88	87	87	87	86	86	86	85	85	
95	93	93	92	92	92	92	91	91	91	90	90	90	90	89	89	89	88	88	88	
99	94	94	94	94	93	93	93	93	92	92	92	92	91	91	91	91	90	90	90	

P_s

P_t	41	42	43	44	45	46	47	48	49	50	51	52	53	54	55	56	57	58	59	60
45	54	53	52	51	50															
50	58	57	57	56	55	54	53	52	51	50										
55	62	62	61	60	59	58	57	57	56	55	54	53	52	51	50					
60	66	65	65	64	63	62	62	61	60	59	58	58	57	56	55	54	53	52	51	50
65	69	69	68	68	67	66	65	65	64	63	63	62	61	60	59	59	58	57	56	55
70	73	72	72	71	70	70	69	68	68	67	67	66	65	64	64	63	62	61	61	60
75	76	75	75	74	74	73	73	72	71	71	70	70	69	68	68	67	67	66	65	64
80	79	78	78	77	77	76	76	75	75	74	74	73	73	72	72	71	71	70	69	69
85	82	81	81	80	80	80	79	79	78	78	77	77	76	76	76	75	75	74	74	73
90	84	84	84	83	83	83	82	82	82	81	81	80	80	80	79	79	78	78	78	77
95	87	87	87	86	86	86	85	85	85	84	84	84	83	83	83	82	82	82	81	8
99	89	89	89	88	88	88	88	87	87	87	87	86	86	86	86	85	85	85	85	84

P_s

P_t	61	62	63	64	65	66	67	68	69	70	71	72	73	74	75	76	77	78	79	80
55	54	53	52	51	50															
60	59	58	57	56	55	54	53	52	51	50										
65	64	63	62	61	61	60	59	58	57	56	55	54	53	51	50					
70	68	68	67	66	65	65	64	63	62	62	61	60	59	58	57	55	54	53	52	5
75	72	72	71	71	70	70	69	68	68	67	66	65	65	64	63	62	61	60	59	5
80	77	76	76	75	75	74	74	73	73	72	72	71	71	70	69	69	68	67	66	6
85	81	80	80	80	79	79	79	78	78	77	77	77	76	76	75	75	74	74	73	7
89	84	84	83	83	83	83	82	82	82	82	81	81	81	80	80	80	80	79	79	7

P_s

P_t	81	82	83	84	85	86	87	88	89	90	91	92	93	94	94	96	97	98	99
75	56	55	54	52	50														
80	64	63	62	61	60	58	57	55	53	50									
85	72	72	71	70	69	68	68	66	65	64	62	60	58	54	50				
89	78	78	78	77	77	77	76	76	75	75	74	74	73	72	71	70	67	63	50

229

Q

II. TABLE OF $2\arcsin\sqrt{P(A)}$

P(A)	0·000	0·001	0·002	0·003	0·004	0·005	0·006	0·007	0·008	0·00
0·00	0·000	0·063	0·090	0·110	0·127	0·142	0·155	0·168	0·179	0·19
0·01	0·200	0·210	0·220	0·229	0·237	0·246	0·254	0·262	0·269	0·27
0·02	0·284	0·291	0·298	0·305	0·311	0·318	0·324	0·330	0·336	0·34
0·03	0·348	0·354	0·360	0·365	0·371	0·376	0·382	0·387	0·392	0·39
0·04	0·403	0·403	0·413	0·418	0·423	0·428	0·432	0·437	0·442	0·44
0·05	0·451	0·456	0·460	0·465	0·469	0·474	0·478	0·482	0·487	0·49
0·06	0·495	0·499	0·503	0·508	0·512	0·520	0·520	0·524	0·528	0·53
0·07	0·536	0·540	0·543	0·547	0·551	0·555	0·559	0·562	0·566	0·57
0·08	0·574	0·577	0·581	0·585	0·588	0·592	0·595	0·599	0·602	0·60
0·09	0·609	0·613	0·616	0·620	0·623	0·627	0·630	0·634	0·637	0·64
0·10	0·644	0·647	0·650	0·654	0·657	0·660	0·663	0·667	0·670	0·67
0·11	0·676	0·679	0·683	0·686	0·689	0·692	0·695	0·698	0·701	0·70
0·12	0·708	0·711	0·714	0·717	0·720	0·723	0·726	0·729	0·732	0·73
0·13	0·738	0·741	0·744	0·747	0·750	0·753	0·755	0·758	0·761	0·76
0·14	0·767	0·770	0·773	0·776	0·779	0·781	0·784	0·787	0·790	0·79
0·15	0·795	0·798	0·801	0·804	0·807	0·809	0·812	0·815	0·818	0·82
0·16	0·823	0·826	0·829	0·831	0·834	0·837	0·839	0·842	0·845	0·84
0·17	0·850	0·853	0·855	0·858	0·861	0·863	0·866	0·869	0·871	0·87
0·18	0·876	0·879	0·882	0·884	0·887	0·889	0·892	0·894	0·897	0·90
0·19	0·902	0·905	0·907	0·910	0·912	0·915	0·917	0·920	0·922	0·92
0·20	0·927	0·930	0·932	0·935	0·937	0·940	0·942	0·945	0·947	0·95
0·21	0·952	0·955	0·957	0·960	0·962	0·964	0·967	0·969	0·972	0·97
0·22	0·977	0·979	0·981	0·984	0·986	0·989	0·991	0·993	0·996	0·99
0·23	1·000	1·003	1·005	1·008	1·010	1·012	1·015	1·017	1·019	1·02
0·24	1·024	1·026	1·029	1·031	1·033	1·036	1·038	1·040	1·043	1·04
0·25	1·047	1·050	1·052	1·054	1·057	1·059	1·061	1·063	1·066	1·06
0·26	1·070	1·073	1·075	1·077	1·079	1·082	1·084	1·086	1·088	1·09
0·27	1·093	1·095	1·097	1·100	1·102	1·104	1·106	1·109	1·111	1·11
0·28	1·115	1·118	1·120	1·122	1·124	1·126	1·129	1·131	1·133	1·13
0·29	1·137	1·140	1·142	1·144	1·146	1·148	1·151	1·153	1·155	1·15
0·30	1·159	1·162	1·164	1·166	1·168	1·170	1·172	1·175	1·177	1·17
0·31	1·181	1·183	1·185	1·188	1·190	1·192	1·194	1·196	1·198	1·20
0·32	1·203	1·205	1·207	1·209	1·211	1·213	1·215	1·218	1·220	1·22
0·33	1·224	1·226	1·228	1·230	1·232	1·235	1·237	1·239	1·241	1·24
0·34	1·245	1·247	1·249	1·251	1·254	1·256	1·258	1·260	1·262	1·26
0·35	1·266	1·268	1·270	1·272	1·275	1·277	1·279	1·281	1·283	1·28
1·36	1·287	1·289	1·291	1·293	1·295	1·298	1·300	1·302	1·304	1·30
0·37	1·308	1·310	1·312	1·314	1·316	1·318	1·320	1·322	1·324	1·32
0·38	1·329	1·331	1·333	1·335	1·337	1·339	1·341	1·343	1·345	1·34
0·39	1·349	1·351	1·353	1·355	1·357	1·359	1·361	1·363	1·365	1·36

P(A)	0·000	0·001	0·002	0·003	0·004	0·005	0·006	0·007	0·008	0·009
0·40	1·370	1·372	1·374	1·376	1·378	1·380	1·382	1·384	1·386	1·388
0·41	1·390	1·392	1·394	1·396	1·398	1·400	1·402	1·404	1·406	1·408
0·42	1·410	1·412	1·414	1·416	1·418	1·420	1·422	1·424	1·426	1·428
0·43	1·430	1·432	1·434	1·436	1·439	1·441	1·443	1·445	1·447	1·449
0·44	1·451	1·453	1·455	1·457	1·459	1·461	1·463	1·465	1·467	1·469
0·45	1·471	1·473	1·475	1·477	1·479	1·481	1·483	1·485	1·487	1·489
0·46	1·491	1·493	1·495	1·497	1·499	1·501	1·503	1·505	1·507	1·509
0·47	1·511	1·513	1·515	1·517	1·519	1·521	1·523	1·525	1·527	1·529
0·48	1·531	1·533	1·535	1·537	1·539	1·541	1·543	1·545	1·547	1·549
0·49	1·551	1·553	1·555	1·557	1·559	1·561	1·563	1·565	1·567	1·569
0·50	1·571	1·573	1·575	1·577	1·579	1·581	1·583	1·585	1·587	1·589
0·51	1·591	1·593	1·595	1·597	1·599	1·601	1·603	1·605	1·607	1·609
0·52	1·611	1·613	1·615	1·617	1·619	1·621	1·623	1·625	1·627	1·629
0·53	1·631	1·633	1·635	1·637	1·639	1·641	1·643	1·645	1·647	1·649
0·54	1·651	1·653	1·655	1·657	1·659	1·661	1·663	1·665	1·667	1·669
0·55	1·671	1·673	1·675	1·677	1·679	1·681	1·683	1·685	1·687	1·689
0·56	1·691	1·693	1·695	1·697	1·699	1·701	1·703	1·705	1·707	1·709
0·57	1·711	1·713	1·715	1·717	1·719	1·721	1·723	1·726	1·728	1·730
0·58	1·732	1·734	1·736	1·738	1·740	1·742	1·744	1·746	1·748	1·750
0·59	1·752	1·754	1·756	1·758	1·760	1·762	1·764	1·766	1·768	1·770
0·60	1·772	1·774	1·776	1·778	1·780	1·782	1·785	1·787	1·789	1·791
0·61	1·793	1·795	1·797	1·799	1·801	1·803	1·805	1·807	1·809	1·811
0·62	1·813	1·815	1·817	1·819	1·822	1·824	1·826	1·828	1·830	1·832
0·63	1·834	1·836	1·838	1·840	1·842	1·844	1·846	1·848	1·851	1·853
0·64	1·855	1·857	1·859	1·861	1·863	1·865	1·867	1·869	1·871	1·873
0·65	1·876	1·878	1·880	1·882	1·884	1·886	1·888	1·890	1·892	1·893
0·66	1·897	1·899	1·901	1·903	1·905	1·907	1·909	1·911	1·914	1·916
0·67	1·918	1·920	1·922	1·924	1·926	1·928	1·931	1·933	1·935	1·937
0·68	1·939	1·941	1·943	1·946	1·948	1·950	1·952	1·954	1·956	1·959
0·69	1·961	1·963	1·965	1·967	1·969	1·972	1·974	1·976	1·978	1·980
0·70	1·982	1·985	1·987	1·989	1·991	1·993	1·996	1·998	2·000	2·002
0·71	2·004	2·007	2·009	2·011	2·013	2·015	2·018	2·020	2·022	2·024
0·72	2·026	2·029	2·031	2·033	2·035	2·038	2·040	2·042	2·044	2·047
0·73	2·049	2·051	2·053	2·056	2·058	2·060	2·062	2·065	2·067	2·069
0·74	2·072	2·074	2·076	2·078	2·081	2·083	2·085	2·088	2·090	2·092
0·75	2·094	2·097	2·099	2·101	2·104	2·106	2·108	2·111	2·113	2·115
0·76	2·118	2·120	2·122	2·125	2·127	2·129	2·132	2·134	2·137	2·139
0·77	2·141	2·144	2·146	2·148	2·151	2·153	2·156	2·158	2·160	2·163
0·78	2·165	2·168	2·170	2·173	2·175	2·177	2·180	2·182	2·185	2·187
0·79	2·190	2·192	2·195	2·197	2·199	2·202	2·204	2·207	2·209	2·211

P(A)	0·000	0·001	0·002	0·003	0·004	0·005	0·006	0·007	0·008	0·00
0·80	2·214	2·217	2·219	2·222	2·224	2·227	2·229	2·232	2·235	2·23
0·81	2·240	2·242	2·245	2·247	2·250	2·252	2·255	2·258	2·260	2·26
0·82	2·265	2·268	2·271	2·273	2·276	2·278	2·281	2·284	2·286	2·28
0·83	2·292	2·294	2·297	2·300	2·302	2·305	2·308	2·311	2·313	2·31
0·84	2·319	2·321	2·324	2·327	2·330	2·332	2·335	2·338	2·341	2·34
0·85	2·346	2·349	2·352	2·355	2·358	2·360	2·363	2·366	2·369	2·37
0·86	2·375	2·378	2·380	2·383	2·386	2·389	2·392	2·395	2·398	2·40
0·87	2·404	2·407	2·410	2·413	2·416	2·419	2·422	2·425	2·428	2·43
0·88	2·434	2·437	2·440	2·443	2·447	2·450	2·453	2·456	2·459	2·46
0·89	2·466	2·469	2·472	2·475	2·478	2·482	2·485	2·488	2·492	2·49
0·90	2·498	2·502	2·505	2·508	2·512	2·515	2·518	2·522	2·525	2·52
0·91	2·532	2·536	2·539	2·543	2·546	2·550	2·554	2·557	2·561	2·56
0·92	2·568	2·572	2·576	2·579	2·583	2·587	2·591	2·595	2·598	2·60
0·93	2·606	2·610	2·614	2·618	2·622	2·626	2·630	2·634	2·638	2·64
0·94	2·647	2·651	2·655	2·660	2·664	2·668	2·673	2·677	2·682	2·68
0·95	2·691	2·695	2·700	2·705	2·709	2·714	2·719	2·724	2·729	2·73
0·96	2·739	2·744	2·749	2·755	2·760	2·765	2·771	2·776	2·782	2·78
0·97	2·794	2·799	2·805	2·812	2·818	2·824	2·831	2·837	2·844	2·85
0·98	2·858	2·865	2·873	2·880	2·888	2·896	2·904	2·913	2·922	2·93
0·99	2·941	2·952	2·963	2·974	2·987	3·000	3·015	3·032	3·052	3·07
1·00	3·142									

232

III. Z-TO-P CONVERSION

Z	P	Z	P	Z	P	Z	P
0·000	0·500	1·000	0·841	2·000	0·977	3·000	0·999
0·050	0·520	1·050	0·853	2·050	0·980	3·050	0·999
0·100	0·540	1·100	0·864	2·100	0·982	3·100	0·999
0·150	0·560	1·150	0·875	2·150	0·984	3·150	0·999
0·200	0·579	1·200	0·885	2·200	0·986	3·200	0·999
0·250	0·599	1·250	0·894	2·250	0·988	3·250	1·000
0·300	0·618	1·300	0·903	2·300	0·989	3·300	1·000
0·350	0·637	1·350	0·912	2·350	0·991	3·350	1·000
0·400	0·656	1·400	0·919	2·400	0·992	3·400	1·000
0·450	0·674	1·450	0·927	2·450	0·993	3·450	1·000
0·500	0·692	1·500	0·933	2·500	0·994	3·500	1·000
0·550	0·709	1·550	0·940	2·550	0·995	3·550	1·000
0·600	0·726	1·600	0·945	2·600	0·995	3·600	1·000
0·650	0·742	1·650	0·951	2·650	0·996	3·650	1·000
0·700	0·758	1·700	0·956	2·700	0·997	3·700	1·000
0·750	0·773	1·750	0·960	2·750	0·997	3·750	1·000
0·800	0·788	1·800	0·964	2·800	0·998	3·800	1·000
0·850	0·802	1·850	0·968	2·850	0·998	3·850	1·000
0·900	0·816	1·900	0·971	2·900	0·998	3·900	1·000
0·950	0·829	1·950	0·975	2·950	0·999	3·950	1·000

233

IV. *P*-TO-*z* CONVERSION AND ORDINATES OF A
STANDARD NORMAL DISTRIBUTION

P	Z	Y	1−P
0·001	3·090	0·003	0·999
0·002	2·878	0·006	0·998
0·003	2·748	0·009	0·997
0·004	2·652	0·012	0·996
0·005	2·576	0·015	0·995
0·010	2·326	0·027	0·990
0·020	2·054	0·049	0·980
0·030	1·881	0·068	0·970
0·040	1·751	0·086	0·960
0·050	1·645	0·103	0·950
0·060	1·555	0·119	0·940
0·070	1·476	0·134	0·930
0·080	1·405	0·149	0·920
0·090	1·341	0·162	0·910
0·100	1·282	0·176	0·900
0·110	1·227	0·188	0·890
0·120	1·175	0·200	0·880
0·130	1·126	0·212	0·870
0·140	1·080	0·223	0·860
0·150	1·037	0·233	0·850
0·160	0·995	0·243	0·840
0·170	0·954	0·253	0·830
0·180	0·915	0·262	0·820
0·190	0·878	0·271	0·810

234

P	Z	Y	1 − P
0·200	0·842	0·280	0·800
0·210	0·807	0·288	0·790
0·220	0·772	0·296	0·780
0·230	0·739	0·304	0·770
0·240	0·706	0·311	0·760
0·250	0·675	0·318	0·750
0·260	0·643	0·324	0·740
0·270	0·613	0·331	0·730
0·280	0·583	0·337	0·720
0·290	0·553	0·342	0·710
0·300	0·525	0·348	0·700
0·310	0·496	0·353	0·690
0·320	0·468	0·358	0·680
0·330	0·440	0·362	0·670
0·340	0·413	0·367	0·660
0·350	0·385	0·370	0·650
0·360	0·359	0·374	0·640
0·370	0·332	0·378	0·630
0·380	0·306	0·381	0·620
0·390	0·279	0·384	0·610
0·400	0·253	0·386	0·600
0·410	0·228	0·389	0·590
0·420	0·202	0·391	0·580
0·430	0·176	0·393	0·570
0·440	0·151	0·395	0·560
0·450	0·126	0·396	0·550
0·460	0·101	0·397	0·540
0·470	0·075	0·398	0·530
0·480	0·050	0·399	0·520
0·490	0·025	0·399	0·510
0·500	0·000	0·399	0·500

235

This page intentionally left blank

INDEX

237

T - #0525 - 101024 - C0 - 229/152/13 - PB - 9780805853230 - Gloss Lamination